INEQUALITY THEORY AND APPLICATIONS
VOLUME 4

INEQUALITY THEORY AND APPLICATIONS

EDITORS: YEOL JE CHO, JONG KYU KIM AND SEVER S. DRAGOMIR

Inequality Theory and Applications. Volume 4
2006. ISBN 1-59454-873-0

Inequality Theory and Applications Volume 3
2004. ISBN 1-59033-891-X

Inequality Theory and Applications. Volume 2
2003. ISBN 11-59033-866-9

Inequality Theory and Applications. Volume 1
2002. ISBN 1-59033-188-5

INEQUALITY THEORY AND APPLICATIONS VOLUME 4

YEOL JE CHO
JONG KYU KIM
SEVER S. DRAGOMIR
EDITORS

Nova Science Publishers, Inc.
New York

For permission to use material from this book please contact us:
Telephone 631-231-7269; Fax 631-231-8175
Web Site: http://www.novapublishers.com

Library of Congress Cataloging-in-Publication Data

ISBN 13 978-1-59454-874-1
ISBN 10 1-59454-874-9

Published by Nova Science Publishers, Inc. ✦New York

CONTENTS

PREFACE

On August 9~13, 2004, "the 8th International Conference on Nonlinear Functional Analysis and Applications" was held at Gyeongsang National University (GNU) and Kyungnam University (KU) by the financial support of the Research Institute of Natural Sciences of GNU, the Institute of Basic Science Research of KU, the Korean Mathematical Society, Nova Science Publishers, Inc., New York, and others.

The aim of the conference is to introduce and exchange recent new topics on the areas of fixed point theory, convex and set-valued analysis, variational inequality and complementarity problem theory, nonlinear ergodic theory, difference, differential and integral equations, control and optimization theory, dynamic system theory, inequality theory, stochastic analysis and probability theory, and their applications.

We have published the following books as the series in mathematical analysis and applications. These books consist of the reviewed papers by the referees, which were given at the 8th Conference, and some invited papers of leading scholars in their fields as special contributors:

(∗) **Fixed Point Theory and Applications**
(Editors: Yeol Je Cho, Jong Kyu Kim, Shin Min Kang),
(∗) **Differential Equations and Applications**
(Editors: Yeol Je Cho, Jong Kyu Kim, Ki Sik Ha),
(∗) **Inequality Theory and Applications**
(Editors: Yeol Je Cho, Jong Kyu Kim, Sever S. Dragomir).

We are sure that these books will contribute and spur further research in the topics mentioned above. We are grateful and thankful to all the authors for their contribution to these books. These books would not have been published without their cooperation, support and help.

Finally, we would like to thank Dr. Frank Columbus, President of Nova Science Publishers, Inc., New York, and all the staffs for publishing the beautiful books.

July 20, 2005

YEOL JE CHO (Gyeongsang National University, Korea)
JONG KYU KIM (Kyungnam University, Korea)
SEVER S. DRAGOMIR (Victoria University of Technology, Australia)

In: Inequality Theory and Applications, Volume 4
Editors: Yeol Je Cho et al., pp. 1–9
ISBN 1-59454-874-9

INSTABILITY AND SUPER-ADDITIVITY

TRANDAFIR BĂLAN[1] AND MARIA PREDOI[1]

ABSTRACT. We analyze a variant of instability, in which instead of non-continui-ty of the function *initial state-evolution*, we propose the dual of continuity, name-ly *discreteness*, in the sense of the *horistological* structures. We show that a dual of the classical Lyapunov function can be defined, and the second method is applicable to study the discrete instability of dynamical systems in $\mathbb{R}^n$.

1. Introduction

Even if restricted to the Lyapunov's sense, the notion of stability has plenty of meanings. For example, the standard form of the stability of a linear system reduces to the boundedness of the evolutions, while for non-linear smooth systems it is expressed by the continuity of the map *initial state-evolution*. Many variants have been produced by altering the space of functions that stand for evolutions, or by stressing on some particular components of the evolution. The common feature of all these variants, which was therefore assumed in the most general system theory, is the continuity of this particular function. More exactly (see [8], etc.), the evolution of a dynamical system is defined as a function $x : T \to \mathcal{S}$, where $T \subseteq \mathbb{R}$ is the *time set*, and $\mathcal{S}$ is the set of *states*. Most frequently, we take $T = [t_0, \to)$ for some $t_0 \in \mathbb{R}$, where t_0 is referred to as *initial moment*. Correspondingly, $x_0 = x(t_0)$ is called *initial state*. The subsequent changes of state are described by an internal rule of state transition, $\lambda : \mathcal{S} \times K \to \mathcal{S}$, where K is the usual order of T (i.e., induced from $\mathbb{R}$) and $y = \lambda(x, (t_1, t_2))$ means that the state x at the moment t_1 is transformed into the state y at the later moment t_2. Such a dynamical system is said to be of *time evolution* type and it is shortly noted by the triplet $(\mathcal{S}, T, \lambda)$.

Generalizing the case of some smooth systems, described by differential equations with unique solutions of the Cauchy's problem (see [1], [7], etc.), it is always assumed that these evolutions are uniquely determined by the initial states via the internal rule of state transition. This correspondence, say $\Psi : \mathcal{S} \to \mathcal{S}^T$, is known as *initial state - evolution* function, and the notation $x = \Psi(x_0)$ shows that the evolution x starts with the initial value x_0, i.e., $x_0 \stackrel{\Psi}{\longmapsto} x(t, x_0)$. Consequently, we get another way to specify the dynamical system, namely by the triplet $(\mathcal{S}, T, \Psi)$.

The notion of stability involves particular structures on $\mathcal{S}$, e.g. those of a normed linear space. The norm is used to produce topologies on $\mathcal{S}$ and $\mathcal{S}^T$, which are necessary to formulate the condition of continuity. Let us also recall that, using linearity, we may always reduce

Received January 24, 2004.
2000 *Mathematics Subject Classification.* 93D05, 47B50
Key Words and Phrases. Instability, Lyapunov function, horistology, super-additivity, discreteness.
[1] Department of Applied Mathematics, University of Craiova, Str. Al. I. Cuza 13, Romania. E-mails: tbalan@central.ucv.ro and mpred@central.ucv.ro

the problem of stability to the null solution θ, where $\theta(t) = 0$ at all $t \geq t_0$. More exactly, the null evolution $\theta \in \mathcal{S}^T$ is said to be *stable* iff Ψ is continuous at $0 = \theta(t_0) \in \mathcal{S}$, i.e.,

$$\forall_{\varepsilon>0} \ \exists_{\delta>0} \ \forall_{\|x_0\|<\delta} \ \forall_{t\geq t_0} \Longrightarrow \|\Psi(x_0)(t)\| < \varepsilon.$$

As far as we know, people use to reduce the notion of *instability* to NON-continuity and most frequently to the exact negation of the condition from above. In this paper, we aim to show that instability may be conceived as a property dual to continuity, which does not contradict this one. In our opinion, the dual of continuity is *discreteness* in the sense defined in [3], i.e., in a framework specific to super-additivity and its forthcoming structures.

2. Discreteness

Similarly to the continuous functions, which are the morphisms of the topological structures, the *discrete* functions represent the morphisms of the *horistological* spaces. The horistologies (introduced in [3] and shortly presented in [6], [8], etc.) have been conceived as qualitative structures of super-additivity, and they are clearly dual to topologies. In other words, this sense of discreteness is dual to continuity since horistology is dual to topology. This duality is visible in the very starting definitions, where the *filters* of neighborhoods are replaced by *ideals* of perspectives. For the sake of self-containedness, we remember that a horistology on $\mathcal{S}$ is defined by a function $\chi : \mathcal{S} \to \mathcal{P}(\mathcal{P}(\mathcal{S}))$, which attaches a family $\chi(e) \subset \mathcal{P}(\mathcal{S})$, of *perspectives,* to each $e \in \mathcal{S}$, such that the following conditions hold:

[h_1] $\forall_{e\in\mathcal{S}} \ \forall_{P\in\chi(e)} \Longrightarrow e \notin P$;
[h_2] $P, Q \in \chi(e) \Longrightarrow P \cup Q \in \chi(e)$;
[h_3] $P \in \chi(e), \ Q \subset P \Longrightarrow Q \in \chi(e)$;
[h_4] $\forall_{P\in\chi(e)} \ \exists_{W\in\chi(e)} \ \forall_{\nu\in P} \ \forall_{Q\in\chi(\nu)} \Longrightarrow Q \subseteq W$.

The pair $(\mathcal{S},\chi)$ is called *horistological* space.

It is significant to recall that each horistological space is endowed with a *proper order* defined by

$$\Pi_\chi = \{(e, \nu) \in \mathcal{S}^2 : \{\nu\} \in \chi(e)\} \cup \Delta,$$

where $\Delta = \{(e, e) : e \in \mathcal{S}\}$ is the *diagonal* of $\mathcal{S}^2$, and represents the equality on $\mathcal{S}$.

If $(\mathcal{S}, \chi)$ and $(\mathcal{T}, \xi)$ are two horistological spaces, then a function $f : \mathcal{S} \to \mathcal{T}$ is said to be *discrete* at $e \in \mathcal{S}$ iff $P \in \chi(e)$ always implies $f(P) \in \xi(f(e))$. If f is discrete at each $e \in D \subseteq \mathcal{S}$, we say that f is discrete *on* D. It is easy to see that in this case f is monotonic on D relative to the proper orders of the horistologies χ and ξ. It is also remarkable that differently from continuous functions, for which the counter-images of the $f(x)-$ neighborhoods are asked to be $x-$ neighborhoods, the discrete functions directly carry $e-$ perspectives into $f(e)-$ perspectives (similarly to boundedness, Darboux property, etc.).

3. Metric Horistologies

Now we have to explain how $\mathbb{R}$, $\mathbb{R}^n$, etc., can be viewed as discrete (i.e., horistological) spaces. For explicitness, we recall that the sub-additivity (briefly, s.a.) of a metric is referred to as the usual inequality in triangles. By duality, i.e. reversing this inequality, we will consider the *super-additive metrics*:

Definition 1. A function $\rho : K \to \mathbb{R}_+$, where K is an order relation on $\mathcal{S}$, is named *super-additive* (briefly S.a.) metric on $\mathcal{S}$ iff the following conditions are fulfilled:

[M_1] $\rho(x, y) = 0$ iff $x = y$ (non-degeneracy),
[M_2] $\rho(x, y) \geq \rho(x, z) + \rho(z, y)$ *whenever* $(x, z), (z, y) \in K$ (super-additivity).

The pair $(\mathcal{S}, \rho)$ is called S.a. *metric space.*

Obviously, the S.a. metrics cannot be defined on the whole $\mathcal{S}^2$; their most natural domains are the orders of $\mathcal{S}$.

In linear spaces, we may similarly speak of *super-additive norms*, which are introduced by the following:

Definition 2. Let $\mathcal{S}$ be a real linear space and let $K \subset \mathcal{S}^2$ be an order on $\mathcal{S}$, which is compatible with linearity. The set $K[0] = \{e \in \mathcal{S} : (0, e) \in K\}$ denotes the *order cone of vertex* 0, which corresponds to K. A function $\dashv \cdot \vdash : K[0] \to \mathbb{R}_+$ is called *super-additive* (briefly, S.a.) norm on $\mathcal{S}$ iff it satisfies the conditions:

$[N_1]$ $\dashv x \vdash = 0$ iff $x = 0$;
$[N_2]$ $\dashv \lambda x \vdash = \lambda \dashv x \vdash$ *for all* $\lambda \geq 0$ *and* $x \in K[0]$;
$[N_3]$ $\dashv x + y \vdash \geq \dashv x \vdash + \dashv y \vdash$ *for all* $x, y \in K[0]$.

The pair $(\mathcal{S}, \dashv \cdot \vdash)$ is called S.a. *normed space.*

Similarly to the sub-additive norms and metrics, it is easy to prove that:

Proposition 1. *If* $(\mathcal{S}, \dashv \cdot \vdash)$ *is a S.a. normed space, where* $\dashv \cdot \vdash : K[0] \to \mathbb{R}_+$, *then the function* $\rho : K \to \mathbb{R}_+$, *of values*

$$\rho(x, y) = \dashv y - x \vdash$$

is a S.a. metric on $\mathcal{S}$.

In [2], there are enough examples to conclude that all sub-additive norms/metrics have super-additive pairs. Therefore the topological and the horistological structures on the same set generally can be coupled. In fact, super-additivity leads to horistology in a way similar to the construction of the metric topology. More exactly, if K is an order on $\mathcal{S}$ and $\rho : K \to \mathbb{R}_+$ is a super-additive metric, then we may take $P \subset \mathcal{S}$ to be a perspective of $e \in \mathcal{S}$ iff there is some $\gamma > 0$ such that $P \subseteq H(e, \gamma)$, where $H(e, \gamma) = \{\nu \in \mathcal{S} : \rho(e, \nu) > \gamma\}$ is the hyperbolic perspective of e, which has the *radius* γ. It is significant to mention that $K = \Pi_\chi$ whenever χ is generated by the S.a. metric $\rho : K \to \mathbb{R}_+$.

Because the study of the practical problems usually involves measurements, the metric horistologies will be of primary interest in the context of (in)stability too. The following examples of S.a. metrics and corresponding horistologies will be particularly used to study the instability.

Example 1. We take $\mathcal{S} = \mathbb{R}$, K equals the natural order of $\mathbb{R}$, and $\rho(x, y) = y - x$ at any $(x, y) \in K$. Consequently, $P \subset \mathcal{S}$ is a perspective of $x \in \mathbb{R}$ iff $P \subseteq [y, +\infty)$ holds for some $y > x$. The resulting horistology of $\mathbb{R}$ is qualified as *natural.*

A similar construction is possible towards $-\infty$.

Example 2. $\mathcal{S} = \mathbb{R}^2$, K is the product order, i.e.,

$$K = \{((x, y), (u, v) : x < u \ \textit{and} \ y < v\} \cup \Delta,$$

and ρ is the hyperbolic metric of values

$$\rho((x, y), (u, v) = (u - x)(v - y).$$

Example 3. $\mathcal{S} = \mathbb{R}^2$, K is the product order and $\delta((x, y), (u, v)) = \min\{u - x, v - y\}$.

Example 4. $\mathcal{S} = \mathbb{R}^n$, where $n \in \mathbb{N} \setminus \{0, 1, 2\}$, K is the product order and the S.a. metrics are similar to ρ and δ from above.

Example 5. $\mathcal{S} = \mathbb{R}^T, \Lambda$ is the product order of functions, i.e.

$$\Lambda = \{(x,y) \in \mathcal{S}^2 : \underset{\eta > 0}{\exists} \underset{t \in T}{\forall} \Longrightarrow x(t) + \eta < y(t)\} \cup \Delta,$$

and $\sigma(x,y) = \inf\{y(t) - x(t) : t \in T\}$.

It is easy to identify the sub-additive metrics in duality with the examples from above. We may remark that the dual metrics are not uniquely determined. For example, the Euclidean metric of $\mathbb{R}^n$, where $n \in \mathbb{N}$, can be considered in duality with both $\mu : \Pi \to \mathbb{R}_+$ and $\vartheta : K \to \mathbb{R}_+$ of values

$$\mu((x_1, x_2, ..., x_n), (y_1, y_2, ..., y_n)) = \left(\sum_{k=1}^{n} \sqrt{y_k - x_k} \right)^2,$$

where $((x_1, x_2, ..., x_n), (y_1, y_2, ..., y_n) \in \Pi$ iff, at all $k = \overline{1,n}$, we have either $x_k = y_k$ or $x_k < y_k$ and, respectively,

$$\vartheta((t, x_1, x_2, ..., x_n), (\tau, y_1, y_2, ..., y_n)) = \left[c^2(\tau - t)^2 - \sum_{k=1}^{n} (y_k - x_k)^2 \right]^{\frac{1}{2}},$$

where $c > 0$, and the *events* $(t, x_1, x_2, ..., x_n)$ and $(\tau, y_1, y_2, ..., y_n)$ are in *causal* order K (according to the physical significance of the elements of this example in the special relativity framework (see [10], [5], etc.)).

4. Instability

Definition 3. The null evolution $\theta \in \mathcal{S}^T$ of a dynamical system $(\mathcal{S}, T, \Psi)$ is said to be *discretely instable* (briefly, d.i.) iff Ψ is discrete at the initial state $0 \in \mathcal{S}$.

Because discreteness depends on the particular horistologies, say χ on $\mathcal{S}$, and ξ on $\mathcal{S}^T$, we may be more specific by mentioning that Ψ is $\chi - \xi$ discrete, and the system is $\chi - \xi$ d.i. As long as no relation between topologies and horistologies, including duality, is a priori imposed on individual spaces, stability and discrete instability are in principle independent properties of the dynamical systems. Therefore we may expect them to hold in particular cases simultaneously (as in Example 6 below).

The mathematical pendulum is generally agreed as a standard example in the dichotomy *stability - instability,* since the normal pendulum is stable, but the reversed one is not. The following proposition shows that the reversed pendulum is a very natural example of d.i. system too.

Proposition 2. *The reversed free unhitted pendulum is discretely instable.*

Proof. The free pendulum of length l, in the gravitational field of acceleration g, evolves according to the equation

$$\varphi'' - \omega^2 \sin \varphi = 0,$$

where $\omega^2 = g/l$. The unknown function φ represents the angle between the rod and the vertical through the fixed end of the rod, directed upwards. The value of the mass $m > 0$, carried at the other end of the rod, does not influence the evolution. This equation is non-linear and difficult to solve, but we may change the unknown function and introduce $\varsigma(\varphi) = \varphi'(t)$. By integrating the resulting equation

$$\varsigma' \varsigma - \omega^2 \sin \varphi = 0$$

we obtain the relation

$$\varsigma^2(\varphi) = -2\omega^2 \cos\varphi + C.$$

Because there is no impulse on the pendulum, we have $\varphi'(0) = 0$, hence the initial state is determined only by $\varphi(0) = \varphi_0$. Consequently, we may express the constant C by φ_0 and we obtain

$$\varsigma^2(\varphi) = 2\omega^2[\cos\varphi_0 - \cos\varphi].$$

To keep up the evolution in the real range, we have to restrict $\varphi \in [\varphi_0, 2\pi - \varphi_0]$, so that $\cos\varphi \leq \cos\varphi_0$.

It is easy to see that $\varphi_1(t) = 0$ and $\varphi_2(t) = \pi$ are stationary evolutions of the pendulum, which correspond to the initial states 0, and respectively π. To show that φ_1 is a d.i. evolution, we remark that the conditions of the above definition are fulfilled for $\mathcal{S} = [0, 2\pi] \subset \mathbb{R}$, K and ρ as in the above example 4, $T = \mathbb{R}_+$, i.e. $t_0 = 0$, and finally, Λ and σ from the example 8 on $\mathcal{S}^T \cap C^2(T)$. More exactly, the *initial state - evolution* function Ψ, for which $\varphi_1 = \Psi(0)$, is discrete at 0, relative to the horistologies generated by ρ and σ on $\mathcal{S}$, respectively $\mathcal{S}^T$.

Another classical case in the problem of stability/instability is that of the time invariant linear systems. A remarkable theorem establishes that such a system is (internally) stable iff the proper values of the main matrix all have negative real parts, and those with null real parts are simple. Taking into account the form of the fundamental matrix, we may immediately reduce stability to boundedness of some exponential functions. Relative to d.i. we have:

Proposition 3. *A time invariant linear system*

$$x' = Ax + Bu$$

is d.i. iff the proper values of A are all real and strictly positive.

Proof. Because A and B are constant matrices, the evolutions of such a system have the form

$$x(t) = e^{At}\, x_0 + \int_0^t e^{A(t-\tau)} B\, u(\tau)\, d\tau,$$

where $x : \mathbb{R}_+ \to \mathbb{R}^n$, $n \in \mathbb{N}^*$, represents a smooth vector function. The discreteness of Ψ at 0 relative to the product horistology on $\mathcal{S}$ and $\mathcal{S}^T$(as in the above examples 7 and 8) asks each component of x to be an increasing exponential. This means that each proper value of A is strictly positive.

This proposition shows that the discrete instability of time invariant linear systems is more restrictive than non-stability. However, the concomitance of stability and d.i. is possible in spite of their opposite nature, as in the following:

Example 6. Let us consider the second order linear differential equation

$$x'' \cosh t + 2x' \sinh t = 0,$$

where $t \geq t_0 = 0$, and $x \in \mathcal{S} = C^2_{\mathbb{R}}(T)$. If we note $x' = y$, then the equation becomes

$$y' + 2y \tanh t = 0,$$

which has the immediate solution $y(t) = C_1 \cosh^{-2} t$. Finally, the general solution of the starting equation is

$$x(t) = C_1 \tanh t + C_2,$$

where $C_2 = x(0)$, and $C_1 = x^{/}(0) = y(0)$. Obviously, the null evolution corresponds to the null initial conditions $x(0) = x^{/}(0) = 0$. According to the boundedness of the function tanh, for any other evolution we have

$$C_2 \leq x(t) \leq C_1 + C_2,$$

at each $t \geq 0$. Consequently, the *initial state-evolution* function Ψ is both continuous and discrete at the initial state $(x(0), x'(0)) \in \mathbb{R}^2$ relative to the horistology in the examples 4 and 6.

5. Changing the Structures

Because d.i. clearly depends on the horistological structures the involved spaces are endowed with, it is interesting to analyze how this property is affected by the change of structures. Here we will consider only the process of passing to *symmetric* horistology, and to *comparable* horistologies.

Definition 4. If $\rho : K \to \mathbb{R}_+$ is a S.a. metric on $\mathcal{S}$, then $\rho^* : K^{-1} \to \mathbb{R}_+$, where $\rho^*(x, y) = \rho(y, x)$, is said to be *symmetric* to ρ.

Obviously, ρ^* is a S.a. metric too. If ρ generates the horistology χ, then ρ^* generates another horisology, say χ^*, which is naturally considered *symmetric* to χ . Alternatively, instead of speaking of two S.a. metrics ρ and ρ^*, we may reformulate the condition $[M_2]$ for a single symmetric S.a. metric. Generally speaking, the existence of a symmetric horistology turns out to be strongly dependent on uniformizability and metrizability (see [4], [8], etc.).
Because the metrical horistologies always appear in pairs, the property of discreteness, and in particular the discrete instability, can be studied in pairs too. For example, the study of the pendulum can be similarly developed for $\varphi \in [-2\pi, 0]$ and the result is the same. Unlike this example, reversing time in linear systems generally affects the physical meaning, and the property of being d.i. is no longer preserved (e.g., see Proposition 3 from above).

Definition 5. Let χ and χ^- be horistologies on the same set $\mathcal{S}$. We say that χ^- is *smaller* than χ iff $\chi^-(e) \subseteq \chi(e)$ holds at every $e \in \mathcal{S}$ and we note $\chi^- \subseteq \chi$. Similarly we define a greater horistology, which is noted $\chi^+ \supseteq \chi$. The horistologies χ^-, χ , and χ^+ are said to be *comparable*.

Passing to comparable horistologies sometimes preserves d.i.:

Proposition 4. *Let $(\mathcal{S}, T, \Psi)$ be a $\chi - \xi$ discretely instable system. If $\chi^- \subseteq \chi$, and $\xi^+ \supseteq \xi$, then the system $(\mathcal{S}, T, \Psi)$ is also $\chi^- - \xi^+$ discretely instable.*

Proof. If Ψ is $\chi - \xi$ discrete, then it is $\chi^- - \xi^+$ discrete too.

6. The Second Method

From now on, we consider smooth dynamical systems in $\mathcal{S} = \mathbb{R}^n$, $n \in \mathbb{N}^*$, which are described by the differential equations

$$x' = f(x), \tag{1}$$

where f is supposed to have enough properties (e.g., Lipschitz) such that $x(0) = x_0$ uniquely determines the evolution $x : T = \mathbb{R}_+ \to \mathbb{R}^n$. In addition, we assume that $f(0) = 0$, which allows $x = 0$ to be a solution of (1). Whenever we speak of (in)stability, we refer to this solution.
The second method of studying the stability reduces the problem to the existence of a so called *Lyapunov function*:

Definition 6. We say that $L : V \to \mathbb{R}$ is a Lyapunov function of (1), where V is a neighborhood of the origin in $\mathbb{R}^n$, iff it satisfies the conditions:

[L_1] *L is continuous at $0 \in \mathbb{R}^n$;*
[L_2] *L is positively definite on V;*
[L_3] *Function $L^* : \mathbb{R}_+ \to \mathbb{R}$ is decreasing, where its values are defined by*

$$L^*(t) = L(x(t, x_0)) = L(\Psi(x_0)(t)).$$

A remarkable theorem states that if a Lyapunov function exists, then the system is stable (see [1], [7], etc.). We remember that its proof is based on the existence of some particular functions called *scale functions.*

Our aim is to show that the same method can be applied to study the discrete instability of (1). The key of the problem is the construction of an adequate Lyapunov type function. In this respect we specify that $\mathbb{R}$ will be endowed with its natural horistology, and $\mathcal{S} = \mathbb{R}^n$ will be organized as a S.a. normed space using an arbitrary S.a. norm $\dashv \cdot \vdash : K[0] \to \mathbb{R}_+$, where K is a linear order on $\mathbb{R}^n$. We note by χ the horistology generated by $\dashv \cdot \vdash$. Finally, $\mathcal{S}^T$ will be endowed with the horistology ξ, generated by the S.a. norm

$$\dashv\mid x \mid\vdash = \inf\{\dashv x(t) \vdash : t \geq 0\}.$$

Consequently, the $\chi - \xi$ discrete instability of (1), i.e., the discreteness of Ψ at 0, means that

$$\forall_{\delta > 0} \exists_{\gamma > 0} \text{ such that } \dashv x_0 \vdash \geq \delta \Longrightarrow \dashv\mid x \mid\vdash \geq \gamma .$$

7. Discrete Lyapunov Functions

In order for us to introduce the discrete counterpart of L, we have first to specify a particular class of functions, which allows to construct the horistological analogous of the scale functions of the classical Lyapunov's theory.

Definition 7. Let $\mathcal{D}$ be the class of functions $D : K[0] \to \mathbb{R}_+$, for which:

[D_1] D is discrete on $K[0]$ relative to χ and to the natural horistology of $\mathbb{R}_+$;
[D_2] D is *gradually bounded,* i.e. for every $r > 0$ there is some $x(r) \in K[0]$, with $\dashv x(r) \vdash \leq r$, such that

$$D(x(r)) = \max\{D(x) : x \in K[0], \dashv x \vdash \leq r\};$$

[D_3] $D(0) = 0 = \inf\{D(x(r)) : r > 0\}$.

If $D \in \mathcal{D}$, then the function $\omega_D : \mathbb{R}_+ \to \mathbb{R}_+$ of values

$$\omega_D(r) = D(x(r))$$

is called *hyperbolic scale* attached to D.

Example 7. The function $D = k \dashv \cdot \vdash$, where $k > 0$ is fixed, is of class $\mathcal{D}$. The corresponding hyperbolic scale is the linear function $\omega_D(r) = kr$.

Proposition 5. *Let us consider a function $E : K[0] \to \mathbb{R}_+$ and for each $r \in \mathbb{R}_+$, let us define the r-slice by $Sl(r) = \{x \in K[0] : \dashv x \vdash \leq r\}$. Then:*

(a) *If E satisfies $[D_1]$, and $[D_3]$, then it is positively definite on $K[0]$;*
(b) *E satisfies condition $[D_2]$ iff it is bounded on each r - slice and reaches its supremum at some $x(r) \in Sl(r)$;*
(c) *If $D, E \in \mathcal{D}$, then $\alpha D + \beta E \in \mathcal{D}$ for all $\alpha, \beta > 0$.*

Proof. (a) For each $x \in K[0] \setminus \{0\}$, there is some $\delta > 0$ such that $\dashv x \vdash > \delta$ and hence $\{x\}$ is a perspective of $0 \in K[0]$. Because E is discrete at 0, to this δ there corresponds some $\gamma > 0$, such that $E(x) > \gamma$. Consequently, E is strictly positive on $K[0] \setminus \{0\}$. In addition, $E(0) = 0$ follows from the first part of $[D_3]$.

(b) We may rewrite $[D_2]$ in the form: for every $r > 0$ there is some $x(r) \in Sl(r)$ such that

$$E(x(r)) = \sup\{E(x) : x \in Sl(r)\}.$$

(c) $\alpha D + \beta E$ fulfils the conditions $[D_1] - [D_3]$.

Lemma 6. *For all $D \in \mathcal{D}$, the hyperbolic scale ω_D has the following properties:*

(i) *$\omega_D(\dashv x \vdash) \geq D(x)$ at every $x \in K[0]$;*
(ii) *ω_D is strictly increasing;*
(iii) *$\omega_D(0) = 0 = \inf\{\omega_D(r) : r > 0\}$.*

Proof. (i) For each $x \in K[0]$, we note $r = \dashv x \vdash$ and we remark that $x \in Sl(r)$. In particular, if $x = 0$, then $r = 0$, and $Sl(0) = \{0\}$.

(ii) We have to show that $0 \leq r_1 < r_2$ implies $\omega_D(r_1) < \omega_D(r_2)$. In fact, let $x_1 = x(r_1) \in Sl(r_1)$ be a point, where

$$D(x_1) = \max\{D(x) : x \in Sl(r_1)\}.$$

The discreteness of D at x_1 means that, for every $\delta > 0$, we can find an $\gamma > 0$ such that the inequality $D(x) \geq D(x_1) + \gamma$ holds whenever $\dashv x \vdash \geq \delta$. In other words, $\dashv x \vdash \geq \delta$ implies $\omega_D(x) \geq \omega_D(x_1) + \gamma$. It remains to take δ such that $H(x_1, \delta) \cap Sl(r_2) \neq \emptyset$.

(iii) This is a simple reformulation of $[D_3]$.

Corollary 7. *The hyperbolic scales are discrete relative to the natural horistology.*

Proof. In the natural horistology of $\mathbb{R}$, we have $\dashv x \vdash = x$ at any $x \in K[0] = \mathbb{R}_+$ and hence the discreteness of ω_D is a simple consequence of (ii).

Now, we may define the discrete variant of Lyapunov function:

Definition 8. Let $D^* : \mathbb{R}_+ \to \mathbb{R}_+$ be a function attached to the pair consisting of an evolution $x(t, x_0)$ of (1) and a function $D \in \mathcal{D}$ according to the formula

$$D^*(t) = D(x(t, x_0)).$$

We say that D is a discrete Lyapunov function of (1) iff D^* is increasing.

Example 8. If we put $n = 1$ and $A = (a)$ in (1), then each initial state $x_0 \geq 0$ determines an evolution of the form

$$x(t, x_0) = x_0 \ \exp(at).$$

If $D = k \dashv \cdot \vdash$, as before, for some $k > 0$, then

$$D^*(t) = k \ x(t, x_0) = k \ x_0 \ \exp(at).$$

Obviously, $D \in \mathcal{D}$. On the other hand, if $a > 0$, then the function D^* is increasing, hence D is a discrete Lyapunov function of the particular system (1).

Theorem 8. *If the system* (1) *allows a discrete Lyapunov function, then its null evolution is discretely instable.*

Proof. According to the above Lemma 6(i), we have

$$\omega_D(\dashv x(t,x_0)\vdash) \geq D(x(t,x_0)). \tag{2}$$

Because D^* is increasing, by Definition 8, it follows that

$$D(x(t,x_0)) = D^*(t) \geq D^*(0) = D(x_0). \tag{3}$$

Finally, since D is discrete, for every $\delta > 0$, there exists an $\varepsilon > 0$ such that $\dashv x_0 \vdash \geq \delta$ implies

$$D(x_0) \geq \varepsilon. \tag{4}$$

By transitivity, from (2), (3), and (4), we deduce that

$$\omega_D(\dashv x(t,x_0)\vdash) \geq \varepsilon \tag{5}$$

holds as a consequence of $\dashv x_0 \vdash \geq \delta$. In accordance to the formula (iii) of Lemma 6, let $\gamma > 0$ be a point where $\omega_D(\gamma) \leq \varepsilon$ holds. Consequently, from (5), we deduce the inequality

$$\omega_D(\dashv x(t,x_0)\vdash) \geq \omega_D(\gamma). \tag{6}$$

Since ω_D is strictly increasing, the inequality (6) leads to similar inequality between arguments, i.e.,

$$\dashv x(t,x_0)\vdash \geq \gamma, \tag{7}$$

holds at arbitrary $t > 0$. In terms of $\dashv\| \cdot \|\vdash$, applied to $x(t,x_0) = \Psi(x_0)(t)$, (7) becomes $\dashv\| \Psi(x_0) \|\vdash \geq \gamma$. To conclude, we can say that, for arbitrary $\delta > 0$, there exists $\gamma > 0$ such that $\dashv x_0 \vdash \geq \delta$ implies $\dashv\| \Psi(x_0) \|\vdash \geq \gamma$, i.e., Ψ is discrete at 0. This completes the proof.

Using this theorem, we can immediately prove that the particular system (1), which we have considered in Example 8 from above, is discretely instable.

REFERENCES

1. V. Arnold, *Équations Différentielles Ordinaires*, Ed. Mir, Moscou, 1978.
2. T. Bălan, *Generalizing the Minkowskian Space-Time (I)*, Studii Cerc. Mat. (Mathematical Reports) **44** (1992), 89–107.
3. T. Bălan, *Generalizing the Minkowskian Space-Time (II)*, Studii Cerc. Mat. (Mathematical Reports) **44** (1992), 267–284.
4. T. Bălan, *Symmetric Horistologies*, Proc. Scientific Workshop on Topological Linear Ordered Spaces, Sinaia, 1992, Nr.13 (1992).
5. Bălan T., Chiriac I., *Relativity in Real Time*, (in Romanian), Ed. Universitaria, Craiova, 2001.
6. T. Bălan and M. Predoi, *Aczél's Inequality, Super-additivity and Horistology*, Inequality Theory and Applications, Vol. 2, Eds.: Y. J. Cho, J. K. Kim and S. S. Dragomir, Nova Science Publishers, New York, 2003.
7. W. E. Boyce and T. C. Di Prima, *Elementary Differential equations and Boundary Value Problems*, 4th edition, Wiley, New York, 1986.
8. I. Chiriac, *Qualitative Structures of Super-additivity*, Preprint Series in Mathematics, West University of Timişoara **No. 79** (1997), 1–25.
9. M. D. Mesarovič and Y. Takahara, *General Systems Theory - Mathematical Foundations*, Accad. Press, New York, San-Francisco, London, 1975.
10. L. G. Naber, *The Geometry of Minkowski Space-time*, Springer-Verlag, Berlin, Heidelberg, New York, 1992.

In: Inequality Theory and Applications, Volume 4
Editors: Yeol Je Cho et al., pp. 11–14
ISBN 1-59454-874-9

SOME REMARKS ON THE NOISELESS CODING THEOREM

N. S. Barnett[1] and S. S. Dragomir[1]

Abstract. An improvement of the Noiseless Coding Theorem for certain probability distributions is given.

1. Introduction

The following analytic inequality for the $\log(\cdot)$ map is well known in the literature (see for example [1, Lemma 1.2.2, p. 22]):

Lemma 1. *Let $P = (p_1, \cdots, p_n)$ be a probability distribution that is, $0 \le p_i \le 1$ and $\sum_{i=1}^n p_i = 1$. Let $Q = (q_1, \cdots, q_n)$ have the property that $0 \le q_i \le 1$ and $\sum_{i=1}^n q_i \le 1$, then*

$$\sum_{i=1}^n p_i \log_b \frac{1}{p_i} \le \sum_{i=1}^n p_i \log_b \frac{1}{q_i}, \quad b > 1, \tag{1.1}$$

where $0 \log_b \frac{1}{0} = 0$ and $p \log_b \frac{1}{0} = +\infty$ for $p > 0$. Furthermore, the equality holds if and only if $q_i = p_i$ for all i.

Note that the proof of this result in [1] uses the elementary inequality:

$$\ln x \le x - 1, \quad x > 0.$$

We give here an alternative proof based on the concavity of the mapping $\log_r(\cdot)$.

As the mapping $f(x) = \log_r(x)$ $(r > 1)$ is a strictly concave mapping on $(0, \infty)$, we have

$$f(x) - f(y) \ge f'(x)(x - y), \quad x, y > 0,$$

i.e., as $f'(x) = \frac{1}{\ln r} \cdot \frac{1}{x}$ for $x > 0$,

$$\log_r x - \log_r y \ge \frac{1}{\ln r}\left(\frac{x - y}{x}\right), \quad x, y > 0. \tag{1.2}$$

Choosing $x = \frac{1}{q_i}$, $y = \frac{1}{p_i}$ in (1.2) gives

$$\log_r \frac{1}{q_i} - \log_r \frac{1}{p_i} \ge \frac{1}{\ln r}\left(\frac{p_i - q_i}{p_i}\right), \quad i \in \{1, \cdots, n\}. \tag{1.3}$$

Received March 16, 2004.
2000 *Mathematics Subject Classification.* Primary 94 A xx; Secondary 26 D xx.
Key words and phrases. Entropy, Instantaneous Codes, The Noiseless Coding Theorem
[1] School of Computer Science and Mathematics, Victoria University of Technology, PO Box 14428, Melbourne City MC, Victoria 8001, Australia. E-mails: sever@matilda.vu.edu.au and neil@matilda.vu.edu.au

Multiplying this inequality by $p_i > 0$ $(i = 1, \cdots, n)$, we get

$$p_i \log_r \frac{1}{q_i} - p_i \log_r \frac{1}{p_i} \geq \frac{1}{\ln r} (p_i - q_i), \quad i \in \{1, \cdots, n\}.$$

Summing over i from 1 to n gives

$$\sum_{i=1}^{n} p_i \log_r \frac{1}{q_i} - \sum_{i=1}^{n} p_i \log_r \frac{1}{p_i} \geq \frac{1}{\ln r} \left(\sum_{i=1}^{n} p_i - \sum_{i=1}^{n} q_i \right)$$
$$= \frac{1}{\ln r} \left(1 - \sum_{i=1}^{n} q_i \right) \geq 0$$

and the inequality (1.1) is obtained. The case of equality follows by the strict concavity of the mapping $\log_r$.

In this paper, by use of (1.1), we point out an improvement to the Noiseless Coding Theorem.

2. The Main Results

Consider an encoding scheme $(c_1, \cdots, c_n)$ for a probability distribution $(p_1, \cdots, p_n)$. The *average codeword length of an encoding scheme* $(c_1, \cdots, c_n)$ for $(p_1, \cdots, p_n)$ is

$$AveLen (c_1, \cdots, c_n) = \sum_{i=1}^{n} p_i len (c_i).$$

We denote the length $len (c_i)$ by l_i.

The r-ary entropy of a probability distribution is given by

$$H_r (c_1, \cdots, c_n) = \sum_{i=1}^{n} p_i \log_r \left(\frac{1}{p_i} \right).$$

The following theorem is well known in the literature (see for example [1, Theorem 2.3.1, p. 62]):

Theorem 2. *Let $C = (c_1, \cdots, c_n)$ be an instantaneous (or uniquely decipherable) encoding scheme for $P = (p_1, \cdots, p_n)$, then,*

$$H_r (p_1, \cdots, p_n) \leq AveLen (c_1, \cdots, c_n)$$

with equality if and only if $l_i = \log_r \left(\frac{1}{p_i} \right)$ for all $i = 1, \cdots, n$.

The following result, providing a counterpart inequality, holds.

Theorem 3. *Let $P = (p_1, \cdots, p_n)$ be a given probability distribution and $r \in \mathbf{N}$, $r \geq 2$. If $\varepsilon > 0$ is fixed and there exists natural numbers $l_1, \cdots, l_n$ such that:*

$$\log_r \left(\frac{1}{p_i} \right) \leq l_i \leq \log_r \left(\frac{r^{\varepsilon}}{p_i} \right), \quad i \in \{1, \cdots, n\}, \tag{2.1}$$

then there exists an instantaneous r-ary code $C = (c_1, \cdots, c_n)$ with codeword length $len (c_i) = l_i$ such that

$$H_r (p_1, \cdots, p_n) \leq AveLen (c_1, \cdots, c_n) \leq H_r (p_1, \cdots, p_n) + \varepsilon. \tag{2.2}$$

Proof. Note that (2.1) is equivalent to

$$\frac{1}{p_i} \leq r^{l_i} \leq \frac{r^{\varepsilon}}{p_i}, \quad i \in \{1, \cdots, n\}. \tag{2.3}$$

Now, since $\frac{1}{r^{l_i}} \leq p_i \ (i = 1, \cdots, n)$, it follows that

$$\sum_{i=1}^{n} \frac{1}{r^{l_i}} \leq \sum_{i=1}^{n} p_i = 1$$

and, by Kraft's theorem (see for example [1. Theorem 2.1.2, p. 44]), there exists an instantaneous r-ary code $C = (c_1, \cdots, c_n)$ such that $len\,(c_i) = l_i$. Obviously, by Theorem 2, the first inequality in (2.2) holds.

We have

$$AveLen\,(c_1, \cdots, c_n) = \sum_{i=1}^{n} p_i l_i = \sum_{i=1}^{n} p_i \log_r r^{l_i} = \sum_{i=1}^{n} p_i \log_r \frac{1}{q_i}$$

choosing $q_i = \frac{1}{r^{l_i}} \in [0,1]$. Also, by Kraft's theorem, $\sum_{i=1}^{n} q_i \leq 1$.

By Lemma 1, we have

$$\begin{aligned} 0 &\leq \sum_{i=1}^{n} p_i \log_r \frac{1}{q_i} - \sum_{i=1}^{n} p_i \log_r \frac{1}{p_i} = AveLen\,(c_1, \cdots, c_n) - H_r\,(p_1, \cdots, p_n) \\ &= \sum_{i=1}^{n} p_i \left(\log_r r^{l_i} - \log_r \frac{1}{p_i} \right) = \left| \sum_{i=1}^{n} p_i \left(\log_r r^{l_i} - \log_r \frac{1}{p_i} \right) \right| \\ &\leq \sum_{i=1}^{n} p_i \left| l_i - \log_r \left(\frac{1}{p_i} \right) \right| \leq \varepsilon \sum_{i=1}^{n} p_i = \varepsilon \end{aligned}$$

since, by (2.1), $0 \leq l_i - \log_r \frac{1}{p_i} \leq \log_r r^{\varepsilon} = \varepsilon$.

We shall use the notation:

$$MinAveLen_r\,(p_1, \cdots, p_n)$$

to denote the minimum average codeword length among all r-ary instantaneous encoding schemes for the probability distribution $P = (p_1, \cdots, p_n)$.

The following Noiseless Coding Theorem is well known in the literature (see for example [1, Theorem 2.3.2, p. 64]):

Theorem 4. *For any probability distribution* $P = (p_1, \cdots, p_n)$, *we have*

$$H_r\,(p_1, \cdots, p_n) \leq MinAveLen_r\,(p_1, \cdots, p_n) < H_r\,(p_1, \cdots, p_n) + 1. \tag{2.4}$$

The following question is then a natural one to pose.

Question. Is it possible to replace the constant 1in the above inequality by a smaller one $\varepsilon \in (0,1)$ and, if so, under what conditions for the probability distribution $P = (p_1, \cdots, p_n)$?

The following is a partial answer to this question:

Theorem 5. *Let r be a given natural number and $\varepsilon \in (0,1)$. If a probability distribution $P = (p_1, \cdots, p_n)$ satisfies the condition that every closed interval of real numbers*

$$I_i = \left[\log_r \left(\frac{1}{p_i}\right), \log_r \left(\frac{r^{\varepsilon}}{p_i}\right)\right], \quad i \in \{1, \cdots, n\},$$

contains one natural number, then, for that probability distribution P, we have

$$H_r (p_1, \cdots, p_n) \leq MinAveLen_r (p_1, \cdots, p_n) \leq H_r (p_1, \cdots, p_n) + \varepsilon. \tag{2.5}$$

Proof. Suppose that $l_i \in I_i$ $(i = 1, \cdots, n)$ are these natural numbers, then, as above,

$$\sum_{i=1}^{n} \frac{1}{r^{l_i}} \leq \sum_{i=1}^{n} p_i = 1$$

and, by Kraft's theorem, there exists an instantaneous code $C = (c_1, \cdots, c_n)$ such that $len(c_i) = l_i$. For this code, we have (2.1) and, by Theorem 3, the inequality (2.2) for C. Taking the infimum in this inequality over all r-ary instantaneous codes gives (2.5).

Remark 1. The lengths of the intervals I_i are

$$len(I_i) = \log_r \left(\frac{r^{\varepsilon}}{p_i}\right) - \log_r \frac{1}{p_i} = \varepsilon \in (0,1), \quad i = 0, \cdots, n,$$

but we cannot be sure that I_i always contains a natural number. Also, I_i could contain at most one natural number.

The following result can be useful in practice.

Practical Criterion. *Let a_i be n natural numbers, $i = 1, \cdots, n$. If p_i $(i = 1, \cdots, n)$ are such that*

$$\frac{1}{r^{a_i}} \leq p_i \leq \frac{r^{\varepsilon}}{r^{a_i}}, \quad i = 1, \cdots, n, \tag{2.6}$$

and $\sum_{i=1}^{n} p_i = 1$, then there exists an instantaneous code $C = (c_1, \cdots, c_n)$ with $len(c_i) = a_i$ $(i = 1, \cdots, n)$ such that (2.2) holds for the probability distribution $P = (p_1, \cdots, p_n)$.

For other recent results in the applications of Theory of Inequalities in Information Theory and Coding, see the following references.

References

1. S. Roman, *Coding and Information Theory*, Springer-Verlag, New York, Berlin, Heidelberg, 1992.
2. N. M. Dragomir and S. S. Dragomir, *An inequality for logarithms and its application in coding theory*, Indian J. Math. **43(1)** (2001), 13–20.
3. S. S. Dragomir and C. J. Goh, *Some bounds on entropy measures in information theory*, Appl. Math. Lett. **10(3)** (1997), 23–28.
4. S. S. Dragomir and C. J. Goh, *A counterpart of Jensen's discrete inequality for differentiable convex mappings and applications in information theory*, Math. Comput. Modelling **24(2)** (1996), 1–11.
5. M. Matić, C. E. M. Pearce and J. Pečarić, *Improvements of some bounds on entropy measures in information theory*, Math. Inequal. and Appl. **1** (1998), 295–304.
6. S. S. Dragomir, N. M. Dragomir and K. Pranesh, *Some Estimations of Kraft number and related results*, Soochow J. of Math. **24(4)** (1998), 291–296.
7. S. S. Dragomir, M. Sholz and J. Sunde, *Some upper bounds for relative entropy and applications*, Comp. & Math. with Appl. **39** (2000), 91–100.
8. C. Calude and C. Grozea, *Kraft-Chaitin inequality revisited*, J. UCS **5** (1996), 306–310.

In: Inequality Theory and Applications, Volume 4
Editors: Yeol Je Cho et al., pp. 15–31

ISBN 1-59454-874-9

FUNCTIONALS MEASURE AND INF-COMPACTNESS OF INTEGRAL FUNCTIONALS

Abdelhamid Bourass[1], Bouchaib Ferrahi[1] and Nourddin Saidou[1]

Abstract. We give some disjunction type results for measures and a property of domain of functional measure and we applied them to characterize by growth conditions, the inf-compactness property of an integral functional $I_f(\cdot,\cdot) = \int_\Omega f(\cdot,\cdot,\cdot)d\mu$ defined on the product of two integral spaces.

1. Introduction

Inf-compactness and lower semicontinuity (lsc) properties of integral functionals, play a crucial role in several Mathematics problems (Variational and hemi-variational inequalities, optimization of forms, optimal control, Mathematical economics, statistics, et al.). Those properties have been studied by several authors (E. J. Balder [2], A. Bourass, B. Ferrahi, O. Kahlaoui [5], G. Bottaro, P. Oppezi [3], C. Castaing, P. Clausure [7], L. Césari [9], [10], I. Ekeland, R. Temam [11], E. Giner [13], A. D. Ioffé [14], C. Olech [15], [16], [17], Rockafellar [18]) and others.

In this paper, we are interested by the study of inf-compactness properties of an integral functional with two variables defined as follows:

$$I_f(u,v) = \int_\Omega f(\omega, u(\omega), v(\omega))d\mu,$$

where f is an integrand defined on $\Omega \times X \times Y$, Ω is a measurable space and X, Y are Banach spaces, the functions u and v run over integrals spaces. To our knowledge, it is the first time that the problem of the inf-compactness of an integral functional defined on a product of spaces is studied. In the first paragraph, we present some technical "disjunction type" results for a family of measures and we establish a property of the domain of functional measures where the σ-decomposability plays a key role. In the second paragraph we give growth conditions that insure the inf-compactness of I_{f_u} for all u in a decomposable set and we also give a characterization in order to the integral functional associated to $\inf_{x\in\Gamma(\omega)} f(\omega,x,y)$ be inf-compact. At last, conditions on the integrand f will be introduced in order to obtain the inf-compactness of I_f on all a neighborhood of $u_0 \in L^0_X(\Omega,\mu)$ as soon as it is inf-compact in u_0.

Received November 22, 2003.
2000*Mathematics Subject Classification.* 28A20, 46E30, 49B.
Key words and phrases. Normal integrand, integral functional, inf-compactness, decomposability, σ-decomposability, almost essentially dominated, growth conditions.

[1] Département de Mathématiques et Informatiques, Faculté des Sciences de Rabat, Université Mohamed V–Agdal, B.P. 1014, Rabat, Moroccoia. E-mails: bourass@fsr.ac.ma, ferrahi@ yahoo.com and saidou9@caramail.com

In the sequel, X and Y denote two separable Banach spaces, $\mathcal{B}_X$, $\mathcal{B}_Y$ their borel tribes respectively and $(\Omega, \mathcal{A}, \mu)$ a measurable space such that μ is an atomless complete finite non negative measure. We note $L^0_X(\Omega, \mu)$ (resp. $L^1_X(\Omega, \mu)$) (resp.$L^\infty_X(\Omega, \mu)$) the space of all measurable functions (resp. Bochner integrable functions) (resp. essentially bounded functions) from Ω into X.

Let $f : \Omega \times X \times Y \to \overline{\mathbb{R}}$ be an integrand (i.e., $\mathcal{A} \otimes \mathcal{B}(X) \otimes \mathcal{B}(Y), \mathcal{B}(\overline{\mathbb{R}})$-measurable), which is not everywhere equal to $+\infty$.

Observe that $\omega \to f(\omega, u(\omega), v(\omega))$ is also measurable for every $u \in L^0_X(\Omega, \mu)$ and every $v \in L^0_Y(\Omega, \mu)$. For every $x \in X$, f_x denotes the function defined for each $(w, y) \in \Omega \times Y$ by $f_x(\omega, y) = f(\omega, x, y)$. The conjugate function or polar of f_x with respect to the (Y, Y') duality is defined for each $(\omega, y') \in \Omega \times Y'$ by $f^*_x(\omega, y') = \sup_{y \in Y} (\langle y, y' \rangle - f_x(\omega, y))$. For every $u \in L^0_X(\Omega, \mu)$, set f_u or $f(u, \cdot)$ the function defined for each $(\omega, y) \in \Omega \times Y$ by $f(\omega, u(\omega), y)$, and $f^*_u(\omega, y') = \sup_{y \in Y} (\langle y, y' \rangle - f(\omega, u(\omega), y))$ its polar with respect to the (Y, Y') duality. We define the integral functional associated to f as follows: $I_f : L^1_X(\Omega, \mu) \times L^1_Y(\Omega, \mu) \to \overline{\mathbb{R}}$ with

$$I_f(u, v) = \begin{cases} \int_\Omega f(\omega, u(\omega), v(\omega)) d\mu & \text{if } f(\cdot, u(\cdot), v(\cdot)) \in L^1_{\mathbb{R}}(\Omega, \mu), \\ +\infty & \text{otherwise.} \end{cases}$$

• We say that a function $g : X \times Y \to \overline{\mathbb{R}}$ is $(\|.\|, \sigma)$-lsc if it is lower semicontinuous on $X \times Y$ when X is equipped with the norm topology and Y with the weak topology $\sigma(Y, Y')$. The function g is proper if it takes no $-\infty$ values and not everywhere equal to $+\infty$.

• We say that a function $g : E \to \overline{\mathbb{R}}$ is τ inf-compact, where τ is a topology on E, if the level sets of g (i.e., $\{x \in E, g(x) \leq M\}$ for some scalar M) are compacts with respect to τ.

• For every coercive even non negative normal convex integrand which is continuous and takes zero in the origin (Young function) $\varphi : \Omega \times X \to \overline{\mathbb{R}}_+$, set:

- C_φ: The Orlicz class associated to φ, that is, the domain of I_φ:

$$C_\varphi = \left\{ u \in L^0_X(\Omega, \mu), \int_\Omega \varphi(\omega, u(\omega)) d\mu < +\infty \right\},$$

- L_φ: The vector space generated by C_φ that is $\mathbb{R} C_\varphi$,
- E_φ: The biggest vector subspace contained in C_φ, which exists since $0 \in C_\varphi$,
- $\| \cdot \|_\varphi$: The Gauge of the convex set $\{I_\varphi(\cdot) \leq 1\}$, it defines a norm in L_φ.
- The polar of φ is defined by $\varphi^*(\omega, x') = \sup_{x \in X} (\langle x, x' \rangle - \varphi(\omega, x))$, it is also a Young function,
- We say that φ satisfies a Δ_2-condition if there exists $\lambda > 1$, $K > 0$ and $a(\cdot) \in L^1_{\mathbb{R}+}(\Omega, \mu)$ such that, for almost every $\omega \in \Omega$ and every $x \in X$, we have

$$\varphi(\omega, \lambda x) \leq K \varphi(\omega, x) + a(\omega).$$

• A σ-finite μ-recouvrement of Ω is a non decreasing family $(\Omega_n)_n$ of pairwise disjoint measurable subsets such that $\mu(\Omega \backslash \bigcup_n \Omega_n) = 0$ and $\mu(\Omega_n) \langle +\infty$ for every n (we shall write $\Omega_n \nearrow \Omega$).

• A μ-partition of Ω is a family $(\Omega_n)_n$ of pairwise disjoint measurable subsets such that $\mu(\Omega \backslash \bigcup_n \Omega_n) = 0$.

• We say that a subset D of $L^0_X(\Omega, \mu)$ is decomposable if, for every u_1 and u_2 in D and every measurable subset E, $u_1 \cdot \chi_E + u_2 \cdot \chi_{E^c}$ is in D, where χ_E denotes the characteristic function of E that is $\chi_E(\omega) = 1$ if $\omega \in E$ and $\chi_E(\omega) = 0$ if not. We say that D is σ-decomposable if, for every $(u_n)_n$ in D and every μ-partition $(\Omega_n)_n$ of Ω, the function

$\sum_n u_n \cdot \chi_{\Omega_n}$ is in D. A subset H is rich in D if, for every $u \in D$, there exists a sequence $(u_n)_n$ in H and $\Omega_n \nearrow \Omega$ such that $u_n \cdot \chi_{\Omega_n} = u \cdot \chi_{\Omega_n}$ μ-almost everywhere for every $n \in \mathbb{N}$.

• Let $\Gamma : \Omega \to X \times Y$ be a measurable multiapplication (see [8]), set L^0_Γ its profile (i.e., all $(\mathcal{A}, \mathcal{B}(X) \otimes \mathcal{B}(Y))$-measurable functions $u : \Omega \to X \times Y$ such that $u(\omega) \in \Gamma(\omega)$ μ-almost everywhere).

• Let f and g be two integrands defined on $\Omega \times Y$ into $\overline{\mathbb{R}}$. We say in accordance with [13] that "g is almost essentially dominated by f" if the following condition holds:

$\forall \lambda > 0\ \exists b_\lambda > 0, \exists a_\lambda \in L^1_{\mathbb{R}_+}(\Omega, \mu)$ such that, for μ-almost every ω in Ω,

$$g(\omega, \lambda y) \leq b_\lambda f(\omega, y) + a_\lambda(\omega), \quad y \in Y.$$

If $b_\lambda = 1$ for every $\lambda > 0$, then we say that "g is essentially dominated by f".

2. Disjunction Type Results and Functional Measures

In the following we establish some new disjunction type results. Those results are related to measures, measurable functionals and integrands.

Lemma 1.1. *Let $(\mu_n)_n$ be a sequence of atomless finite non negative measures such that*

$$\mu_n(\Omega) \geq 2^n, \quad n \in \mathbb{N}.$$

Then there exists a subsequence $(\mu_{n_k})_k$ and a family of pairwise disjoint measurable subsets $(A_p)_p$ such that

$$\mu_{n_k}(A_k) = 1, \quad k \in \mathbb{N}.$$

Proof. The construction is based on the following recurrent procedure, we can extract a subsequence $(\mu_{n_k})_k$ such that, for every $p \in \mathbb{N}$, there exists a family $\{A_i, i = 1, \cdots, p\}$ of pairwise disjoint measurable subsets which satisfy

$$\begin{aligned} &\mu_{n_k}(A_k) = 1, \quad k = 1, \cdots, p, \\ &\mu_{n_k}\left(\Omega \backslash \bigcup_{i=1}^{p} A_i\right) \geq 2^{n_k - p}, \quad k = p+1, \cdots. \end{aligned}$$

For $p = 1$, we have $\mu_1(\Omega) \geq 2$ and μ_1 is atomless, then there exists a measurable subset B_1 such that $\mu_1(B_1) = 1$. For every $n \in \mathbb{N}$, we have $\mu_n(\Omega) = \mu_n(B_1) + \mu_n(\Omega \backslash B_1) \geq 2^n$, so $\mu_n(B_1) \geq 2^{n-1}$ or $\mu_n(\Omega \backslash B_1) \geq 2^{n-1}$. Two possibilities present themselves:

(1) We can extract a subsequence $(\mu_{n_k})_k$ such that $\mu_{n_k}(\Omega \backslash B_1) \geq 2^{n_k - 1}$ for every $k \in \mathbb{N}$, here $\mu_{n_1} = \mu_1$. Then we take $A_1 = B_1$ and continue the procedure with $(\mu_{n_k})_k$.

(2) There exists $n_2 \in \mathbb{N}$ such that $\mu_n(B_1) \geq 2^{n-1}$ for every $n \geq n_2$. We have $\mu_1(\Omega \backslash B_1) \geq 1$ and μ_1 is an atomless measure. It is sufficient to choose $A_1 \subset \Omega \backslash B_1$ such that $\mu_1(A_1) = 1$ and continue with the following sequence $(\mu_n)_{1, n_2, n_2+1, \cdots}$.

Thus, in all cases and by noting $(\mu_{n_k})_k$ the extracted subsequence, we have

For $k = 1$,

$$\mu_{n_1}(A_1) = 1.$$

For $k \geq 2$,

$$\begin{aligned} &\mu_{n_k}(\Omega \backslash A_1) \\ &= \begin{cases} \mu_{n_k}(\Omega \backslash B_1) \geq 2^{n_k - 1}, & \text{the first case,} \\ \mu_{n_k}(\Omega \backslash A_1) \geq \mu_{n_k}(B_1) \geq 2^{n_k - 1} & \text{the second case since } n_k \geq n_2, \end{cases} \end{aligned}$$

which completes the construction for p=1. Suppose that we can extract successive subsequence $(\mu_{n_k})_k$ and build a family $\{A_1, \cdots, A_p\}$ of pairwise disjoints measurable subsets such that

$$\mu_{n_k}(A_k) = 1, \quad k = 1, \cdots, p,$$
$$\mu_{n_k}\Big(\Omega \backslash \bigcup_{i=1}^{p} A_i\Big) \geq 2^{n_k - p}, \quad k = p+1, \cdots.$$

Let us show that the property is still true for the following rank p+1. We have $\mu_{n_{p+1}}(\Omega \backslash \bigcup_{i=1}^{p} A_i) \geq 2^{n_{p+1}-p} \geq 2$ and $\mu_{n_{p+1}}$ atomless, there exists $B_{p+1} \subset \Omega \backslash \bigcup_{i=1}^{p} A_i$ such that

$$\mu_{n_{p+1}}(B_{p+1}) = 1.$$

For $k \geq p+2$, we have

$$\mu_{n_k}\Big(\Omega \backslash \bigcup_{i=1}^{p} A_i\Big) = \mu_{n_k}\Big(\Big(\Omega \backslash \bigcup_{i=1}^{p} A_i\Big) \backslash B_{p+1}\Big) + \mu_{n_k}(B_{p+1}) \geq 2^{n_k - p}.$$

Consequently, we have one of the following alternative cases.

(i) $\mu_{n_k}(B_{p+1}) \geq 2^{n_k - p - 1}$,

(ii) $\mu_{n_k}\Big(\Big(\Omega \backslash \bigcup_{i=1}^{p} A_i\Big) \backslash B_{p+1}\Big) \geq 2^{n_k - p - 1}$.

As above, we distinguish two cases:

(1) There exists an infinite number of indices (k) such that

$$\mu_{n_k}\Big(\Big(\Omega \backslash \bigcup_{i=1}^{p} A_i\Big) \backslash B_{p+1}\Big) \geq 2^{n_k - p - 1}$$

and we can extract a subsequence $(\mu_{n_{k_l}})_l$ which verifies

$$\mu_{n_{k_l}}\Big(\Big(\Omega \backslash \bigcup_{i=1}^{p} A_i\Big) \backslash B_{p+1}\Big) \geq 2^{n_{k_l} - p - 1}.$$

So, we set $A_{p+1} = B_{p+1}$ and we continue the procedure with the following subsequence: $\mu_{n_1}, \cdots, \mu_{n_{p+1}}, (\mu_{n_{k_l}})_l$.

(2) There exists $n_{k_0} \in \mathbb{N}$ such that, for every $k \geq k_0$, we have

$$\mu_{n_k}(B_{p+1}) \geq 2^{n_k - p - 1}.$$

Since

$$\begin{aligned}\mu_{n_{p+1}}\Big(\Big(\Omega \backslash \bigcup_{i=1}^{p} A_i\Big) \backslash B_{p+1}\Big) &= \mu_{n_{p+1}}\Big(\Omega \backslash \bigcup_{i=1}^{p} A_i) - \mu_{n_{p+1}}(B_{p+1}\Big) \\ &\geq 2^{n_{p+1} - p} - 1 \geq 1\end{aligned}$$

and the measure $\mu_{n_{p+1}}$ is atomless, then there exists $A_{p+1} \subset (\Omega \backslash \bigcup_{i=1}^{p} A_i) \backslash B_{p+1}$ such that $\mu_{n_{p+1}}(A_{p+1}) = 1$. We continue with the following subsequence:

$$\mu_{n_1}, \cdots, \mu_{n_{p+1}}, (\mu_{n_k})_{k \geq k_0}.$$

Thus, in the two cases, we can build a subsequence $(\mu_{n_l})_l$ and a family $\{A_1, \cdots, A_{p+1}\}$ of pairwise disjoints measurable subsets such that

$$\mu_{n_l}(A_l) = 1, \quad 1 \leq l \leq p+1,$$
$$\mu_{n_l}\Big(\Omega \backslash \bigcup_{i=1}^{p+1} A_i\Big) \geq 2^{n_l - (p+1)}, \quad l \geq p+2.$$

This completes the proof.

Proposition 1.2. *Let $L \subset L^0_X(\Omega, \mu)$ and $M \subset L^0_Y(\Omega, \mu)$ be two vector subspaces such that L is stable by trancature (i.e., $u \cdot \chi_A \in L$ if $u \in L$ and $A \in \mathcal{A}$). Let D be a σ-decomposable subset of L and $F : L \times M \to \overline{\mathbb{R}}$ be a functional such that*

(i) *The application $A \longmapsto F(u \cdot \chi_A, v)$ is σ-additive, i.e.,*

$$F(u \cdot \chi_{\cup_{i=1}^{\infty} A_i}, v) = \sum_{i=1}^{\infty} F(u \cdot \chi A_i, v)$$

for every $(u, v) \in \mathrm{dom}_D F = \{(u, v) \in D \times M,\ F(u, v) < +\infty\}$,

(ii) *There exists $(u_0, v_0) \in D \times M$ such that $F(u_0, v_0) < +\infty$.*

Then

$$\bigcap_{u \in D} \mathrm{dom}\, F(u, \cdot) = \mathrm{dom}\left(\sup_{u \in D} F(u, \cdot)\right).$$

Proof. Without loss of generality, we can suppose that $0 \in D$ and $F(0, 0) = 0$. Indeed, let (u_0, v_0) such that $F(u_0, v_0) < +\infty$. Set $D_0 = D - u_0$ and $F_0(u, v) = F(u + u_0, v + v_0) - F(u_0, v_0)$. We have

(1) $$\bigcap_{u \in D} \mathrm{dom}\, F(u, \cdot) = \bigcap_{u' \in D_0} \mathrm{dom}\, F_0(u', \cdot) + v_0.$$

Indeed, let $v = v_1 - v_0 \in \bigcap_{u \in D} \mathrm{dom}\, F(u, \cdot) - v_0$. For every $u' \in D_0$, there exists $u \in D$ such that $u' = u - u_0$. So

$$\begin{aligned} F_0(u', v) &= F(u' + u_0, v + v_0) - F(u_0, v_0) \\ &= F(u, v_1) - F(u_0, v_0) < +\infty. \end{aligned}$$

Then $\bigcap_{u \in D} \mathrm{dom}\, F(u, \cdot) \subset \bigcap_{u' \in D_0} \mathrm{dom}\, F_0(u', \cdot) + v_0$. For the other inclusion, let $v = v_1 + v_0 \in \bigcap_{u' \in D_0} \mathrm{dom}\, F_0(u', .) + v_0$. For every $u \in D$, there exists $u' \in D_0$ such that $u = u' + u_0$. Therefore,

$$\begin{aligned} F(u, v) &= F(u' + u_0, v_1 + v_0) \\ &= F_0(u', v_1) + F(u_0, v_0) < +\infty. \end{aligned}$$

This implies the desired inclusion. The following equality can be shown by using the same method

(2) $$\mathrm{dom}\left(\sup_{u \in D} F(u, \cdot)\right) = d\omega\left(\sup_{u' \in D_0} F_0(u', \cdot)\right) + v_0.$$

Suppose from now on that $0 \in D$ and $F(0, 0) = 0$. The inclusion

$$\mathrm{dom}\left(\sup_{u \in D} F(u, \cdot)\right) \subset \bigcap_{u \in D} \mathrm{dom}\, F(u, \cdot)$$

is obvious. Let us suppose that the other inclusion does not hold, then there exists $v' \in M$ such that, for every $u \in D$, we have $F(u, v') < +\infty$ and $\sup_{u \in D} F(u, v') = +\infty$. Thus there exists a subsequence $(u_n)_n$ in D such that

$$F(u_n, v') \geq 2^n, \quad n \in \mathbb{N}.$$

For every $A \in \mathcal{A}$, set

$$\mu_n(A) = F(u_n \cdot \chi_A, v').$$

For every $n \in \mathbb{N}$, μ_n is an atomless measure (hypothesis (i)) because it is absolutely continuous by respect to μ. Moreover,

$$\mu_n(\Omega) = F(u_n \cdot \chi_\Omega, v') = F(u_n, v') \geq 2^n, \quad n \in \mathbb{N}.$$

Using Lemma 1.1, there exists a subsequence $(\mu_{n_k})_k$ of $(\mu_n)_n$ and a family $(A_k)_k$ of pairwise disjoints measurable sets such that for every $k \in \mathbb{N}$ we have $\mu_{n_k}(A_k) = 1$. The element $\bar{u} = \sum_k u_{n_k} \cdot \chi_{A_k}$ is in D because it is σ-decomposable. Observing that $\bar{u} \cdot \chi_{A_k} = u_{n_k} \cdot \chi_{A_k}$, we have

$$\begin{aligned} F(\bar{u}, v') &= F\Big(\sum_k u_{n_k} \cdot \chi_{A_k}, v'\Big) = F\Big(\sum_k \bar{u} \cdot \chi_{A_k}, v'\Big) \\ &= F\Big(\bar{u} \cdot \sum_k \chi_{A_k}, v'\Big) = \sum_k F(\bar{u} \cdot \chi_{A_k}, v') \\ &= \sum_k F(u_{n_k} \cdot \chi_{A_k}, v') = \sum_k \mu_{n_k}(A_k) = +\infty, \end{aligned}$$

which contradicts the fact that $F(u, v') < +\infty$ for every $u \in D$. This completes the proof.

As we can show it later, the assertion (i) in Proposition 1.2 can be obtained by using a lower semicontinuity property. Let us consider the following hypothesis.

Let L be a semi-normed vector topological space and τ be its topology such that (L, τ) is topologically decomposable ([13], Definition 1.1.3) and satisfies the following fundamental hypothesis.

$(F.H)$ The application $(u, A) \in L \times \mathcal{A} \longmapsto u \cdot \chi_A \in L$ is separately continuous (it is sufficient to check it at $(0, \emptyset)$), where L is equipped by τ and $\mathcal{A}$ by the so-called Nikodym topology which is defined as follows:

$$\begin{aligned} A_n \xrightarrow{\mathcal{N}} A &\Leftrightarrow \forall B \in \mathcal{A}, \mu(B) < +\infty, \mu(B \cap (A_n \Delta A)) \to 0 \\ &\Leftrightarrow \chi_{A_n} \xrightarrow{L^0_X(\Omega,\mu)} \chi_A. \end{aligned}$$

Example. of topological spaces where the $(F.H)$ holds:

- $L^0_X(\Omega, \mu)$ equipped with the topology of convergence in measure,
- L_φ equipped with the weak topology $\sigma(L_\varphi, L_{\varphi^*})$ when φ is continuous in 0,
- E_φ equipped with the norm $\|.\|_\varphi$ topology when $\bar{\varphi}(\omega, t) := \sup_{\|x\|=t} \varphi(\omega, x)$, is finite.

Lemma 1.3. *Let $F : L \times M \to \overline{\mathbb{R}}$, and D be a τ-closed decomposable subset of L, which contains zero. Suppose that L verifies the fundamental hypothesis $(H.F)$ and, for every $v \in M$, the following conditions hold:*

(i) *The application $A \longmapsto F(u \cdot \chi_A, v)$ is additive for every $(u, v) \in \operatorname{dom} F$ (i.e., for every finite family $(A_i)_{1 \leq i \leq n}$ of pairwise disjoints measurable subsets, we have*

$$F(u \cdot \chi_{\bigcup_{i=1}^n A_i}, v) = \sum_{i=1}^n F(u \cdot \chi_{A_i}, v),$$

(ii) *The functional $F(\cdot, v)$ is proper τ-lsc on D.*

Then the application $A \longmapsto F(u \cdot \chi_A, v)$ is σ-additive for every $(u,v) \in D \times M$.

Proof. Let $v \in M$, $u \in D$ and $(A_n)_n$ be a family of pairwise disjoints measurable subsets. we must check that

$$F(u \cdot \chi_{\bigcup_n A_n}, v) = \lim_{n \to +\infty} \sum_{p=0}^{n} F(u \cdot \chi_{A_p}, v).$$

From (i), it follows that, for every $n \in \mathbb{N}$,

$$F(u \cdot \chi_{\bigcup_n A_n}, v) = \sum_{p=0}^{n} F(u \cdot \chi_{A_p}, v) + F(u \cdot \chi_{B_n}, v),$$

where $B_n = \bigcup_{p>n} A_p$. Note that B_n is a non-increasing sequence and $\bigcap_n B_n = \emptyset$, which implies $B_n \xrightarrow{\mathcal{N}} \emptyset$. We deduce that

$$\tau - \lim_{n \to +\infty} u \cdot \chi_{B_n} = 0, \quad \tau - \lim_{n \to +\infty} \sum_{p=0}^{n} u \cdot \chi_{A_p} = u \cdot \chi_A,$$

where $A = \bigcup_n A_n$. The functional F is proper τ-lsc and so

$$0 = F(0, v) \leq \liminf_{n \to +\infty} F(u \cdot \chi_{B_n}, v)$$

and

$$\begin{aligned} -\infty < F(u \cdot \chi_A, v) &\leq \liminf_{n \to +\infty} F\left(\sum_{p=0}^{n} u \cdot \chi_{A_p}, v\right) \\ &\leq \liminf_{n \to +\infty} \sum_{p=0}^{n} F(u \cdot \chi_{A_p}, v). \end{aligned}$$

It follows that

$$\begin{aligned} \liminf_{n \to +\infty} \sum_{p=0}^{n} F(u \cdot \chi_{A_p}, v) &\geq F(u \cdot \chi_A, v) \\ &\geq \limsup_{n \to +\infty} \sum_{p=0}^{n} F(u \cdot \chi_{A_p}, v) + \liminf_{n \to +\infty} F(u \cdot \chi_{B_n}, v) \\ &\geq \limsup_{n \to +\infty} \sum_{p=0}^{n} F(u \cdot \chi_{A_p}, v), \end{aligned}$$

that is,

$$F(u \cdot \chi_A, v) = \lim_{n \to +\infty} \sum_{p=0}^{n} F(u \cdot \chi_{A_p}, v).$$

Lemma 1.4. *Let X be a separable Banach space and $\varphi : X \longrightarrow \overline{\mathbb{R}}$ be a non-negative function. The following assertions are equivalent:*

(i) *φ is strongly coercive.*
(ii) *φ^{**} is strongly coercive.*

Proof. The implication (ii)⇒(i) is obvious because $\varphi^{**} \leq \varphi$. Suppose that φ is strongly coercive. Then we have

$$\forall \alpha > 0, \ \exists B_\alpha > 0 \text{ such that } \|y\| \geq B_\alpha \Rightarrow \varphi(y) \geq \alpha\|y\|$$

and we have $\sup_{\|y\|\geq B_\alpha} (\alpha\|y\| - \varphi(y)) \leq 0$. From another hand, φ is non negative and so $\sup_{\|y\|\leq B_\alpha} (\alpha\|y\| - \varphi(y)) \leq \alpha B_\alpha$. It follows that for every $y \in Y$, $\varphi(y) \geq \alpha\|y\| - \alpha B_\alpha$. Taking the bipolar, we have $\varphi^{**}(y) \geq \alpha\|y\| - \alpha B_\alpha$. This implies that, for every $\alpha > 0$, we have $\lim_{\|y\|\to+\infty} \frac{\varphi^{**}(y)}{\|y\|} \geq \alpha$.

Lemma 1.5. *Let $(\alpha_n)_n$ be a sequence in $L^0_{\mathbb{R}}(\Omega, \mu)$ and $\alpha \in L^1_{\mathbb{R}^+}(\Omega, \mu)$. If for μ-almost every ω in Ω, we have $\liminf_{n\to+\infty} \alpha_n(\omega) \geq -\alpha(\omega)$, then*

$$\limsup_{n\to+\infty} \int_\Omega \alpha_n^-(\omega) d\mu > -\infty,$$

where $\alpha_n^- = min(\alpha_n, 0)$.

Proof. Let $a \in L^1_{\mathbb{R}^*_+}(\Omega, \mu)$. We have $\liminf_{n\to+\infty} \alpha_n(\omega) \geq -\alpha(\omega) - a(\omega)$ and for every $\omega \in \Omega$, there exists $n_\omega \in \mathbb{N}$ such that $\alpha_n(\omega) > \alpha(\omega) - a(\omega)$ for every $n \geq n_\omega$. Set

$$A_p = \bigcap_{n\geq p} \{\omega \in \Omega : \alpha_n(\omega) > \alpha(\omega) - a(\omega)\}$$

and note that $\Omega = \bigcup_p A_p$. If $(\Omega_p)_p$ is a σ-finite μ-recouvrement of Ω, then the sequence $(\Omega_p \cap A_p)_p$ is still a σ-finite μ-recouvrement of Ω and, for every $n \in \mathbb{N}$, we have

$$\int_\Omega \alpha_n^-(\omega) d\mu = \lim_{p\to+\infty} \int_{\Omega_p\cap A_p} \alpha_n^-(\omega) d\mu.$$

Using a diagonalization lemma due to Attouch [1], we get that there exists an increasing function $\theta : \mathbb{N} \to \mathbb{N}$ such that

$$\limsup_{n\to+\infty} \int_\Omega \alpha_n^-(\omega) d\mu = \limsup_{n\to+\infty} \lim_{p\to+\infty} \int_{\Omega_p\cap A_p} \alpha_n^-(\omega) d\mu \tag{3}$$

$$\geq \limsup_{p\to+\infty} \int_{\Omega_p\cap A_p} \alpha_{\theta(p)}^-(\omega) d\mu. \tag{4}$$

Notice, that for every $p \in \mathbb{N}$ and μ-almost every ω in $\Omega_p \cap A_p$, we have

$$\alpha_{\theta(p)}(\omega) > \alpha(\omega) - a(\omega),$$

which implies

$$\alpha_{\theta(p)}^-(\omega) > \alpha(\omega) - a(\omega).$$

Consequently, we have

$$\begin{aligned} \int_{\Omega_p\cap A_p} \alpha_n^-(\omega) d\mu &> -\int_{\Omega_p\cap A_p} (\alpha(\omega) + a(\omega)) d\mu \\ &> -\int_\Omega (\alpha(\omega) + a(\omega)) d\mu. \end{aligned}$$

It follows that

$$\limsup_{p\to+\infty} \int_{\Omega_p\cap A_p} \alpha_{\theta(p)}^-(\omega) d\mu > -\infty$$

Then we obtain the desired result in virtue of (4). This completes the proof.

In the following result, we give a more general version of Lemma 3.7 of [4].

Lemma 1.6. *Let $(\mu_n)_n$ be a sequence of non negative atomless measures, such that, for every $n \in \mathbb{N}$, $\mu_n(\Omega) = +\infty$. Then there exists a subsequence $(\mu_{n_k})_k$ and a family $(A_k)_k$ of pairwise disjoints measurable subsets such that:*

$$\mu_{n_k}(A_k) = +\infty, \quad k \in \mathbb{N}.$$

Proof. We adapt the proof of ([4], Lemma 3.7) to non absolutely continuous measures.

Lemma 1.7. *Let $\psi : \Omega \times Y \to \mathbb{R}$ be a Young function and $g : \Omega \times Y' \to \mathbb{R}$ be a function. If $v' \in domI_{\psi^*}$ and $A \in \mathcal{A}$ are such that $\int_A g(v')d\mu = +\infty$, then, for every $a > 0$, there exists a measurable subset A' in A with the following property:*

$$\int_{A'} \psi^*(\omega, v'(\omega))d\mu \leq a, \quad \int_{A'} g(\omega, v'(\omega))d\mu = +\infty.$$

Proof. Suppose that the conclusion of the lemma does not hold. Then, for every measurable subset $A' \subset A$, we have one of the following alternative cases:

$$\text{(5)} \qquad \text{(i)} \int_{A'} \psi^*(\omega, v'(\omega))d\mu > a, \qquad \text{(ii)} \int_{A'} g(\omega, v'(\omega))d\mu < +\infty$$

First, we note that $\int_A \psi^*(\omega, v'(\omega))d\mu > a$ because $\int_A g(\omega, v'(\omega))d\mu = +\infty$. Since the measure μ is atomless, there exists a measurable subset A_1 of A such that $\int_{A_1} \psi^*(\omega, v'(\omega))d\mu = a$. Using (5), we will have $\int_{A_1} g(\omega, v'(\omega))d\mu < +\infty$ and, consequently, $\int_{A\backslash A_1} g(v')d\mu = +\infty$. Furthermore, we have

$$\int_A g(\omega, v'(\omega))d\mu = \int_{A\backslash A_1} g(v')d\mu + \int_{A_1} g(\omega, v'(\omega)')d\mu = +\infty,$$

which implies, in virtue of (5), that $\int_{A\backslash A_1} g(\omega, v'(\omega))d\mu = +\infty$. Using the same argument as above, we can say that there exists a measurable subset A_2 of $A\backslash A_1$ such that $\int_{A_2} \psi^*(\omega, v'(\omega))d\mu = a$. Using successively this method we can build a family $(A_k)_k$ of pairwise disjoints measurable subsets such that $\int_{A_k} \psi^*(\omega, v'(\omega))d\mu = a$ for every $k \in \mathbb{N}$. It follows that

$$\int_\Omega \psi^*(\omega, v'(\omega))d\mu \geq \int_{\bigcup_k A_k} \psi^*(\omega, v'(\omega))d\mu = \sum_k \int_{A_k} \psi^*(\omega, v'(\omega))d\mu = +\infty,$$

which contradicts the fact that $v' \in domI_{\psi^*}$. This completes the proof.

2. Inf-Compactness of an Integral Functional

The aim of this section is to study the Inf-compactness property of an integral functional defined on a product of spaces. Note that, for the one variable functionals, there exists several characterizations. We recall the following which is due to Giner ([13], Theorem 5.2.6).

Theorem 2.1. [13] *Let Y be a reflexive separable Banach space, $\psi : \Omega \times Y \to \overline{\mathbb{R}}_+$ be a Young function and $f : \Omega \times Y \to \overline{\mathbb{R}}$ be a normal convex integrand. Consider the following assertions:*

(i) *The functional I_f is proper $\sigma(L_\psi, L_{\psi^*})$ inf-compact,*
(ii) *f^* is almost essentially dominated by ψ^*,*
(iii) *L_{ψ^*} is a subset of $domI_{f^*}$,*
(iv) $\lim_{\|v\|_\psi \to +\infty} \frac{I_f(v)}{\|v\|_\psi} = +\infty.$

Then (i)⇒(ii)⇔(iii)⇔(iv) and, if I_f is not everywhere equal to $+\infty$, then all assertions are equivalent.

When u is fixed in any integral space, the inf-compactness of $I_f(u,\cdot)$ is characterized by the above theorem. All the conditions in this theorem depends on u. In the following, we give an analogous characterization uniformly in u when u run over a decomposable set.

Theorem 2.2. *Let X and Y be two separable Banach spaces with the additional condition that Y is reflexive. Suppose that $\Gamma : \Omega \to X$ is a measurable multiapplication, $\psi : \Omega \times Y \to \overline{\mathbb{R}}$ is a Young function and $f : \Omega \times X \times Y \to \overline{\mathbb{R}}$ is an integrand such that, for μ-almost every ω in Ω,*

(i) *f is $(\|.\|, \sigma)$-lsc on $X \times Y$,*
(ii) *$f(\omega, x, \cdot)$ is convex for every $x \in X$.*

Let $D \subset L^0_X(\Omega,\mu)$ be a rich σ-decomposable subset of L^0_Γ. Let us put

$$g(\omega, y) = \inf_{x\in\Gamma(\omega)} f(\omega, x, y).$$

If, for every $u \in D$, the functional $I_f(u,\cdot)$ is not everywhere equal to $+\infty$, then the following assertions are equivalent:

(a) *$I_f(u,\cdot)$ is proper $\sigma(L_\psi, L_{\psi^*})$ inf-compact for every $u \in D$.*
(b) *I_g is proper $\sigma(L_\psi, L_{\psi^*})$ inf-compact.*
(c) *For every $\lambda > 0$, there exists $b_\lambda > 0$ and $a_\lambda \in L^1_{\mathbb{R}^+}(\Omega,\mu)$ such that*

$$f(\omega, x, y) \geq \psi\left(\omega, \frac{\lambda y}{b_\lambda}\right) + a_\lambda(\omega), \ \ (x, y) \in \Gamma(\omega) \times Y.$$

Proof. (a)⇒(b) Suppose that $I_f(u,.)$ is proper and $\sigma(L_\psi, L_{\psi^*})$ inf-compact for every $u \in D$. Using Theorem 2.1, we have

$$L_{\psi^*} \subset \operatorname{dom} I_{f^*_u}, \quad u \in D. \tag{6}$$

Applying Proposition 1.2 to the functional $F(u, v) = I_{f^*_u}(v)$, we deduce that

$$L_{\psi^*} \subset \operatorname{dom}\left(\sup_{u\in D} I_{f^*_u}\right). \tag{7}$$

But $g(\omega, y) = \inf_{x\in\Gamma(\omega)} f(\omega, x, y)$ and so we have

$$\sup_{u\in D} I_{f^*_u}(v') = I_{g^*}(v'), \quad v' \in L_{\psi^*}. \tag{8}$$

Indeed, let $v' \in L_{\psi^*}$. The inclusion (6) implies that $I_{f^*_u}(v') < +\infty$ for every $u \in L^0_\Gamma$ and, for every $v \in L_\psi$ such that $I_f(u, v) \in \mathbb{R}$, we can write

$$I_{f^*_u}(v') \geq \int_\Omega \langle v'(\omega), v(\omega)\rangle d\mu - \int_\Omega f(\omega, u(\omega), v(\omega)) d\mu > -\infty.$$

So, we have $f^*_u(v') \in L^1_{\mathbb{R}}(\Omega,\mu)$. Regarding to Theorem 1.4 of [6], we have

$$\begin{aligned}
\sup_{u\in D} I_{f^*_u}(v') &= \sup_{u\in D} \int_\Omega \sup_{y\in Y} \left(\langle v'(\omega), y\rangle - f(\omega, u(\omega), y)\right) d\mu \\
&= \int_\Omega \sup_{x\in\Gamma(\omega)} \sup_{y\in Y} \left(\langle v'(\omega), y\rangle - f(\omega, x, y)\right) d\mu \\
&= \int_\Omega \sup_{y\in Y} \Big(\langle v'(\omega), y\rangle - \inf_{x\in\Gamma(\omega)} f(\omega, x, y)\Big) d\mu \\
&= \int_\Omega g^*(\omega, v'(\omega)) d\mu \\
&= I_{g^*}(v')
\end{aligned}$$

and we conclude that I_g is $\sigma(L_\psi, L_{\psi^*})$ inf-compact by virtue of Theorem 2.1.

(b)$\Rightarrow$)a) is obvious since $f(\omega, u(\omega), y) \geq g(\omega, y)$ for every $u \in D$ and every $y \in Y$.

(b)$\Leftrightarrow$(c) by virtue of Theorem 2.1 (ii). This completes the proof.

Corollary 2.3. *With the same hypothesis of Theorem 2.2, if I_{f_u} is $\sigma(L_\psi, L_{\psi^*})$ inf-compact for every u in a rich decomposable subset D of L^0_Γ, then $\bigcup_{u\in D}\{I_{f_u} \leq \lambda\}$ is weakly relatively compact for every $\lambda \in \mathbb{R}$.*

Proof. Under the same notation as above, observe that $\bigcup_{u\in D}\{I_{f_u} \leq \lambda\} \subset \{I_g \leq \lambda\}$ and we conclude with Theorem 2.2.

In the particular case where u runs over the $L^\infty_X(\Omega, \mu)$ space, we obtain the following corollary:

Corollary 2.4. *Let X and Y be two separable Banach spaces, with the additional condition that Y is reflexive. Suppose that $\psi : \Omega \times Y \to \overline{\mathbb{R}}$ is a Young function, $f : \Omega \times X \times Y \to \overline{\mathbb{R}}$ is an integrand with the same hypothesis as in theorem (2.2). Suppose that, for every $u \in L^\infty_X(\Omega, \mu)$, the functional $I_f(u, .)$ is not everywhere equal to $+\infty$. Then, the following assertions are equivalents:*

(i) *For every $u \in L^\infty_X(\Omega, \mu)$, the functional $I_f(u, \cdot)$ is proper $\sigma(L_\psi, L_{\psi^*})$ inf-compact.*
(ii) *For every $M > 0$, the functional I_{f_M} is proper $\sigma(L_\psi, L_{\psi^*})$ inf-compact, where,*

$$f_M(\omega, y) = \inf_{\|x\|\leq M} f(\omega, x, y).$$

Proof. For every $M > 0$, set

$$\Gamma_M(\omega) = \{x \in X : \|x\| \leq M\}.$$

Then $L^0_{\Gamma_M} = B_\infty(0, M)$, where $B_\infty(0, M)$ is the ball centered in 0 with radius M in $L^\infty_X(\Omega, \mu)$. It is sufficient to note that $L^\infty_X(\Omega, \mu) = \bigcup_{M\rangle 0} B_\infty(0, M)$ and to use Theorem 2.2.

The assertion c) in theorem (2.2) implies that the function $g(\omega, y) = \inf_{x\in\Gamma(\omega)} f(\omega, x, y)$ is strongly coercive for μ-almost every ω in Ω. The reciprocal result seems to be true in many situations. More precisely, we have the next proposition.

First, Recall the following technical lemmas which are due to Giner ([13]).

Lemma 2.5. ([13]) *Let φ be a Young function which is continuous in the origin and strongly coercive. Then there exists another Young function ψ which satisfies a Δ_2-condition such that ψ is almost essentially dominated by φ. Moreover, the following topological inclusion hold:*

$$(L_\varphi, \sigma(L_\varphi, E_{\varphi^*})) \hookrightarrow (L_\psi, \sigma(L_\psi, L_{\psi^*})) = (L_\psi, \sigma(L_\psi, L'_\psi)). \tag{9}$$

Lemma 2.6. ([13]) *Let φ be a Young function. Then bounded closed convex subsets of $L_\varphi(\Omega, \mu)$ are exactly those subsets which are $\sigma(L_\varphi, E_{\varphi^*})$-compacts.*

Proposition 2.7. *Let X and Y be two separable Banach spaces, with the additional condition that Y is reflexive. Let $\Gamma : \Omega \to X$ be a measurable multi-application and $f : \Omega \times X \times Y \to \overline{\mathbb{R}}$ be an integrand with the same hypothesis as in Theorem 2.2. Suppose that the function $g : \Omega \times Y \to \overline{\mathbb{R}}$, where $g(\omega, y) = \inf_{x\in\Gamma(\omega)} f(\omega, x, y)$ is strongly coercive for μ-almost every ω in Ω and that $g^*(\cdot, 0) \in L^1_{\mathbb{R}}(\Omega, \mu)$. Then there exists $\psi : \Omega \times Y \to \overline{\mathbb{R}}$*

a Young function which satisfies a Δ_2-condition such that, for every $u \in L^0_\Gamma$, the functional $I_f(u,.)$ is $\sigma(L_\psi, L_{\psi^*})$ inf-compact.

Proof. Set $h(\omega,.) = g(\omega,.) + g^*(\omega,0)$, then h is a non negative strongly coercive function and, by virtue of Lemma 1.4, $h^{**}(\omega,.)$ is also strongly coercive for μ-almost every ω in Ω. Since $g^*(\omega,0)$ is finite almost everywhere, $g^{**}(\omega,.) = h^{**}(\omega,.) - g^*(\omega,.)$ is also strongly coercive. Using Theorem 3.3 of [12], we get that there exists a strongly coercive Young function φ such that

$$g^{**}(\omega,y) \geq \varphi(\omega,y) - g^*(\omega,0), \quad (\omega,y) \in \Omega \times Y.$$

For every $u \in L^0_\Gamma$ and $v \in L_\psi$, we have, by using (2.5), that

$$\begin{aligned} f(\omega,u(\omega),v(\omega)) &\geq g(\omega,v(\omega)) \\ &\geq g^{**}(\omega,v(\omega)) \\ &\geq \varphi(\omega,v(\omega)) - g^*(\omega,0) \end{aligned}$$

for μ-almost ω in Ω. Then we have

$$I_f(u,v) \geq I_\varphi(v) - \int_\Omega g^*(\omega,0)d\mu,$$

which shows that lower sections of $I_f(u,\cdot)$ are bounded in L_φ. Furthermore, the function $f(u(\cdot),\cdot)$ is lsc and bounded below by $-g^*(\cdot,0)$ which is integrable. Thus the lower sections of $I_f(u,\cdot)$ are also closed in measure and convex in L_φ. By using (2.6), we conclude that such sections are $\sigma(L_\varphi, E_{\varphi^*})$-compacts. In virtue of (9) it follows that they are $\sigma(L_\psi, L'_\psi)$-compacts. This completes the proof.

Theorem 2.2 gives inf-compactness characterization of integral functional I_{f_u} for every u in a decomposable set. Now, we give a new inf-compactness condition of I_{f_u} with u in a neighborhood of u_0. In the following definition, we will introduce a new notion of equi-lower semicontinuity in the neighborhood of $+\infty$. The last notion will play an important role in our study.

Definition 2.8. Let X be a topological space and Y be a normed vector space. We say that a function $f : X \times Y \to \overline{\mathbb{R}}$ is *equi-lsc* in the neighborhood of $+\infty$ at $x_0 \in X$ if

$$\begin{aligned} &\forall \epsilon > 0 \;\; \exists V_\epsilon \in \mathcal{V}(x_0) \;\; \exists r_\epsilon > 0 \;\; \text{such that} \\ &\qquad f(x_0,y) \leq f(x,y) + \epsilon, \;\; (x,y) \in V_\epsilon \times Y, \;\; \|y\| > r_\epsilon, \end{aligned}$$

where $\mathcal{V}(x_0)$ denotes the set of all neighborhoods of x_0.

Example. Let $f(\omega,x,y) = p(\omega,x) + q(w,y)$, where $p : \Omega \times X \to \overline{\mathbb{R}}$ and $q : \Omega \times Y \to \overline{\mathbb{R}}$ are functions. If p is lower semicontinuous at x_0, then f is equi-lsc in the neighborhood of $+\infty$ at x_0.

Under this assumption of equi-lsc in the neighborhood of $+\infty$, the strong coercivity of a function in a point implies its "uniform" strong coercivity in a neighborhood of this point:

Lemma 2.9. *Let X be a hausdorff topological space and Y be a normed space, $f : X \times Y \to \overline{\mathbb{R}}$ be a function and $x_0 \in X$. Suppose that*

(i) *f is equi-lsc in the neighborhood of $+\infty$ at x_0,*
(ii) *$f(x_0,\cdot)$ is strongly coercive.*

Then there exists a neighborhood V of x_0 in X such that $\inf_{x\in V} f(x,\cdot)$ is strongly coercive.

Proof. Let $\epsilon > 0$. According to (i), there exists $V \in \mathcal{V}(x_o)$ and $r_\epsilon > 0$ such that $\inf_{x\in V} f(x,y) > f(x_0,y) - \epsilon$ for every $\|y\| > r_\epsilon$. Since $f(x_0,\cdot) - \epsilon$ is strongly coercive, it will be the same for $\inf_{x\in V} f(x,\cdot)$.

Generally, the upper bound of a family of upper semicontinuous functions (usc) is not necessary usc, but, with the assumption of equi-lsc in the neighborhood of $+\infty$, we get back the scs of the following function:

$$x \in X \to f_x^*(y') = \sup_{y\in Y} (\langle y, y'\rangle - f(x,y)) \in \overline{\mathbb{R}}$$

for every fixed $y' \in Y$. Indeed, we have the following result:

Proposition 2.10. *Let X and Y be two separable Banach spaces with the additional hypothesis that Y is reflexive, $f : X \times Y \to \overline{\mathbb{R}}$ be a function which satisfy:*

(i) *f is $(\|.\|,\sigma)$-lsc on $X \times Y$,*
(ii) *f is equi-lsc in the neighborhood of $+\infty$ at x_0.*

Then, for every $y' \in Y'$, the function $x \in X \to f_x^*(y') \in \overline{\mathbb{R}}$ is norm usc at x_0.

Proof. Let $y' \in Y'$ be fixed. We must show that the function:

$$h(x) = -f_x^*(y') = \inf_{y\in Y} (f(x,y) - \langle y, y'\rangle)$$

is lsc at x_0. Let $\epsilon > 0$ and $(x_n)_n \to x_0$ in the norm of X. The hypothesis of equi-lsc in the neighborhood of $+\infty$ implies that there exists $n_1 \in \mathbb{N}$ and $B_\epsilon > 0$ such that for every $n \geq n_1$ and $\|y\| > B_\epsilon$, we have $f(x_n,y) \geq f(x_0,y) - \epsilon$. Then

$$f(x_n,y) - \langle y', y\rangle \geq f(x_0,y) - \langle y', y\rangle - \epsilon \geq h(x_0) - \epsilon$$

for every $\|y\| > B_\epsilon$. It follows that

$$\inf_{\|y\|>B_\epsilon} (f(x_n,y) - \langle y', y\rangle) \geq h(x_0) - \epsilon, \quad n \geq n_1. \tag{10}$$

From another hand, for every $n \in \mathbb{N}$, the function $y \in Y \to f(x_n,y) - \langle y', y\rangle \in \overline{\mathbb{R}}$ reach its minimum on the ball $\{y \in Y, \|y\| \leq B_\epsilon\}$ since this function is weakly lsc and Y is reflexive. Therefore, for every $n \in \mathbb{N}$, there exists $y_n \in Y$, $\|y_n\| \leq B_\varepsilon$ such that

$$\inf_{\|y\|\leq B_\epsilon} (f(x_n,y) - \langle y', y\rangle) = f(x_n,y_n) - \langle y', y_n\rangle. \tag{11}$$

So, we have

$$h(x_0) \leq \liminf_{n\to+\infty} (f(x_n,y_n) - \langle y', y_n\rangle). \tag{12}$$

Indeed, if the second limit is $+\infty$, there is nothing to check. Otherwise, there exists an increasing function $\varphi : \mathbb{N} \to \mathbb{N}$ such that

$$\liminf_{n\to+\infty} (f(x_n,y_n) - \langle y', y_n\rangle) = \lim_{n\to+\infty} \left(f(x_{\varphi(n)}, y_{\varphi(n)} - \langle y', y_{\varphi(n)}\rangle\right) = c < +\infty.$$

The sequence $(y_{\varphi(n)})_n$ is bounded in Y and so we can extract a weakly convergent subsequence $(y_{\psi(\varphi(n))})_n$ which converges to some y_0 in Y. Since the function $f(\cdot,\cdot)- \langle y', \cdot\rangle$ is $(\|\cdot\|,\sigma)$-lsc on $X \times Y$,

$$\begin{aligned} h(x_0) &\leq f(x_0,y_0) - \langle y', y_0\rangle \\ &\leq \liminf_{n\to+\infty} \left(f(x_{\psi(\varphi(n))}, y_{\psi(\varphi(n))}) - \langle y', y_{\psi(\varphi(n))}\rangle\right) = c \end{aligned}$$

and thus (12) is satisfied and there exists $n_2 \in \mathbb{N}$ such that

$$f(x_n, y_n) - \langle y', y_n\rangle \geq h(x_0) - \epsilon, \quad n \geq n_2. \tag{13}$$

By virtue of (10), (11) and (13), there exists an integer $n_0 = \max(n_1, n_2)$ such that

$$h(x_n) = \inf_{y \in Y} (f(x_n, y) - \langle y', y\rangle) \geq h(x_0) - \epsilon, \quad n \geq n_0.$$

This completes the proof of the upper semicontinuity of h and, consequently, $f_x^*(y') = -h(x)$ is lsc at x_0. This completes the proof.

In the following, we check out the fact that, if f is equi-lsc in neighborhood of $+\infty$ and I_f is inf-compact in a point, then the functional $I_f(u, .)$ is inf-compact for every u in a neighborhood of this point.

Theorem 2.11. *Let X and Y be two separable Banach spaces with the additional hypothesis that Y is reflexive. Let $\varphi : \Omega \times X \to \overline{\mathbb{R}}_+$ and $\psi : \Omega \times Y \to \overline{\mathbb{R}}_+$ be two Young functions such that ψ^* satisfies a Δ_2-condition and $f : \Omega \times X \times Y \to \overline{\mathbb{R}}$ be an integrand. Suppose that*

(i) *$f(\omega, \cdot, \cdot)$ is $(\|\cdot\|, \sigma)$-lsc on $X \times Y$ for μ-almost every ω in Ω,*
(ii) *The function $f(\omega, x, \cdot)$ is convex for every $x \in X$ and μ-almost every ω in Ω,*
(iii) *There exists u_0 in L_φ such that, for μ-almost every ω in Ω, the function $f(\omega, \cdot, \cdot)$ is equi-lsc in neighborhood of $+\infty$ at $u_0(\omega)$,*
(iv) *The functional I_f takes no $-\infty$ values,*
(v) *$I_{f_{u_0}}$ is proper $\sigma(L_\psi, L_{\psi^*})$ inf-compact.*

Then there exists a neighborhood V_0 of u_0 in L_φ such that $I_f(u, \cdot)$ is proper $\sigma(L_\psi, L_{\psi^*})$ inf-compact for every $u \in V_0$ such that $I_f(u, \cdot)$ is not equal to $+\infty$.

Proof. In virtue of Theorem 2.1, it is sufficient to check that

$$\begin{aligned} &\exists r > 0 \text{ such that } L_{\psi^*} \subset \operatorname{dom} I_{f_u^*} \text{ for every } u \in B_\varphi(u_0, r) \\ &\qquad \text{such that } \operatorname{dom} I_f(u, \cdot) \neq \emptyset, \end{aligned} \tag{14}$$

where $B_\varphi(u_0, r)$ denotes the ball of L_φ centered in u_0 with radius r. Since the functional $I_f(u_0, .)$ is proper $\sigma(L_\psi, L_{\psi^*})$ inf-compact, Using Theorem 2.1, we get

$$\begin{aligned} &\forall \lambda > 0 \;\; \exists b_\lambda > 0 \;\; \exists a_\lambda \in L^1_{\mathbb{R}^+}(\Omega, \mu) \quad \text{such that} \\ &f^*_{u_0}(\omega, \lambda y') \leq b_\lambda \psi^*(\omega, y') + a(\omega) \end{aligned} \tag{15}$$

for every $y' \in Y'$ and μ-almost every ω in Ω. This is equivalent, for every $y \in Y$ and μ-almost every ω in Ω, to

$$\begin{aligned} &\forall \lambda > 0 \;\; \exists b_\lambda > 0 \;\; \exists a_\lambda \in L^1_{\mathbb{R}^+}(\Omega, \mu) \quad \text{such that} \\ &f(\omega, u_0(\omega), y) \geq b_\lambda \psi\left(\omega, \frac{\lambda y}{b_\lambda}\right) - a_\lambda(\omega). \end{aligned} \tag{16}$$

Suppose that (14) does not hold. Then there would exist two sequences $(u_n)_n$ in L_φ and $(v'_n)_n$ in L_{ψ^*} such that

$$\|u_n - u_0\|_\varphi \leq \frac{1}{n}, \quad I_{f^*_{u_n}}(v'_n) = +\infty, \quad I_f(u_n, \cdot) \neq +\infty. \tag{17}$$

For every $n \in \mathbb{N}^*$, set

$$B_n = \{\omega \in \Omega \mid f^*_{u_n}(\omega, v'_n(\omega)) \le 0\}, \tag{18}$$

$$\bar{u}_n = u_n \cdot \chi_{B_n} + u_0 \cdot \chi_{B_n^c}, \tag{19}$$

$$\bar{v}'_n = v'_n \cdot \chi_{B_n}. \tag{20}$$

The function $f^*_{\bar{u}_n}(\omega, \bar{v}'_n(\omega)) = f^*_{u_n}(\omega, v'_n(\omega)) \cdot \chi_{B_n} + f^*_{u_0}(\omega, 0) \cdot \chi_{B_n^c}$ is well defined because, if $\omega \in \Omega \backslash B_n$, then

$$\begin{aligned} f^*_{\bar{u}_n}(\omega, \bar{v}'_n(\omega)) &= f^*_{u_0}(\omega, 0) \\ &= \sup_{y \in Y} (-f(\omega, u_0(\omega), y)) \\ &\ge -f(\omega, u_0(\omega), v_0(\omega)) > -\infty, \end{aligned}$$

where v_0 is some element of L_ψ such that $I_f(u_0, v_0) \in \mathbb{R}$. From another hand, we have, in virtue of (16),

$$f^*_{\bar{u}_n}(\omega, \bar{v}'_n(\omega)) = \sup_{y \in Y} (-f(\omega, u_0(\omega), y)) \le a_\lambda(\omega) < +\infty.$$

Thus, for every n in $\mathbb{N}$, we have

$$f^*_{\bar{u}_n}(\omega, \bar{v}'_n(\omega)) \cdot \chi_{B_n} \in L^0_{\mathbb{R}_+}(\Omega, \mu), \quad f^*_{u_n}(\cdot, v'_n(\cdot)) \ge 0 \quad \text{on } B_n^c.$$

Then we get

$$\begin{aligned} \int_\Omega f^*_{\bar{u}_n}(\omega, \bar{v}'_n(\omega)) \cdot \chi_{B_n} d\mu &= \int_{B_n} f^*_{u_n}(\omega, v'_n(\omega)) d\mu \\ &= \int_\Omega f^*_{u_n}(\omega, v'_n(\omega)) d\mu - \int_{B_n^c} f^*_{u_n}(\omega, v'_n(\omega)) d\mu \\ &= I_{f^*_{u_n}}(v'_n) - \int_{B_n^c} f^*_{u_n}(\omega, v'_n(\omega)) d\mu \\ &= +\infty. \end{aligned}$$

Applying Lemma 1.6 to the non-negative measures $\mu_n(\cdot) = \int_{(\cdot)} f^*_{\bar{u}_n}(\omega, \bar{v}'_n(\omega)) \cdot \chi_{B_n} d\mu$, we can extract a subsequence $(\bar{u}_{n_k})_k$ and build a family $(A'_k)_k$ of pairwise disjoint measurable subsets such that

$$\int_{A'_k} f^*_{\bar{u}_{n_k}}(\omega, \bar{v}'_{n_k}(\omega)) \cdot \chi_{B_{n_k}} d\mu = +\infty$$

for every $k \in \mathbb{N}$. Set $A_k = A'_k \cap B_{n_k}$, then $(A_k)_{k \in \mathbb{N}}$ is a family of pairwise disjoint measurable subsets such that

$$\int_{A_k} f^*_{u_{n_k}}(\omega, v'_{n_k}(\omega)) d\mu = +\infty, \quad k \in \mathbb{N},$$

but $v'_{n_k} \in L_{\psi^*} = \operatorname{dom} I_{\psi^*}$. Then, for every $k \in \mathbb{N}^*$, Lemma 1.7 insure the existence of a measurable subset $E_k \subset A_k$ such that:

$$\int_{E_k} \psi^*(\omega, v'_{n_k}(\omega)) d\mu \le \frac{1}{2^k}, \quad \int_{E_k} f^*_{u_{n_k}}(\omega, v'_{n_k}(\omega)) d\mu = +\infty. \tag{21}$$

Let us define the following functions:

$$v' = \sum_n v'_{n_k} \cdot \chi_{E_k}, \qquad w_p = u_{n_p} \cdot \chi_{\bigcup_k E_k} + u_o \cdot \chi_{\Omega \setminus \bigcup_k E_k}. \tag{22}$$

Notice that $v' \in L_{\psi^*}$ because

$$\int_\Omega \psi^*(\omega, v'(\omega))d\mu = \sum_k \int_{E_k} \psi^*(\omega, v'(\omega))d\mu \leq \sum_k \frac{1}{2^k} < +\infty$$

and, for every $p \in \mathbb{N}$, we have

$$\begin{aligned} &\int_{\bigcup_{k\neq p} E_k} f^*_{u_{n_p}}(\omega, v'(\omega))d\mu \\ &= \int_{\bigcup_{k\neq p} E_k} \sup_{y\in Y} \left(\langle v'(\omega), y\rangle - f(\omega, u_{n_p}(\omega), y)\right) d\mu \\ &\geq \int_{\bigcup_{k\neq p} E_k} \left(\langle v'(\omega), v_{n_p}(\omega)\rangle - f(\omega, u_{n_p}(\omega), v_{n_p}(\omega))\right) d\mu \\ &\geq -\int_\Omega |\langle v'(\omega), v_{n_p}\rangle| d\mu - \int_\Omega |f(\omega, u_{n_p}(\omega), v_{n_p}(\omega))| d\mu \\ &> -\infty, \end{aligned} \tag{23}$$

where $v_{n_p} \in L_\psi$ is such that $I_f(u_{n_p}, v_{n_p}) \in \mathbb{R}$. This choice is possible because $I_f(\cdot,\cdot)\rangle - \infty$ and $I_f(u_{n_p}, \cdot)$ is not everywhere equal to $+\infty$. For every $p \in \mathbb{N}$, in virtue of (21) and (23), we can write

$$\begin{aligned} &\int_\Omega f^*_{w_p}(\omega, v'(\omega))d\mu \\ &= \int_{\bigcup_k E_k} f^*_{u_{n_p}}(\omega, v'(\omega))d\mu + \int_{\Omega\setminus\bigcup_k E_k} f^*_{u_0}(\omega, v'(\omega))d\mu \\ &\geq \int_{E_p} f^*_{u_{n_p}}(\omega, v'(\omega))d\mu + \int_{\bigcup_{k\neq p} E_k} f^*_{u_{n_p}}(\omega, v'(\omega))d\mu + \int_\Omega |f(\omega, u_0(\omega), v_0(\omega))| d\mu \\ &= +\infty \end{aligned}$$

and so

$$\int_\Omega f^*_{w_p}(\omega, v'(\omega))d\mu = +\infty, \quad p \in \mathbb{N}. \tag{24}$$

From another hand, using (17) and (22), $(w_p)_p$ converges to u_0 in L_φ and, if necessary, take a subsequence, then we can suppose that $(w_p)_p$ converges μ-almost everywhere to u_0. Since, for μ-almost every ω in Ω, the function $x \to f^*_x(\omega, v'(\omega))$ is usc in $u_0(\omega)$ (Proposition 2.10),

$$\limsup_{p\to+\infty} f^*_{w_p(\omega)}(\omega, v'(\omega)) \leq f^*_{u_0(\omega)}(\omega, v'(\omega)) \quad \mu\text{-almost everywhere,}$$

Taking $\lambda = 1$ in (15), there exists $b_1\rangle 0$ and $a_1 \in L^1_{\mathbb{R}_+}(\Omega, \mu)$ such that

$$\limsup_{p\to+\infty} f^*_{w_p(\omega)}(\omega, v'(\omega)) \leq f^*_{u_0(\omega)}(\omega, v'(\omega)) \leq b_1\psi^*(\omega, v'(\omega)) + a_1(\omega),$$

that is,

$$-(b_1\psi^*(\omega, v'(\omega)) + a_1(\omega)) \leq \liminf_{p\to+\infty}(-f^*_{w_p(\omega)}(\omega, v'(\omega))).$$

Finally, notice that $-(b_1\psi^*(\omega, v'(\omega)) + a_1(\omega)) \in L^1_{\mathbb{R}_+}(\Omega, \mu)$ and, in virtue of Lemma 1.5, we obtain

$$-\infty < \limsup_{p\to+\infty} \int_\Omega -f^*_{w_p}(\omega, v'(\omega))d\mu = -\liminf_{p\to+\infty} \int_\Omega f^*_{w_p}(\omega, v'(\omega))d\mu,$$

which contradicts (24). This completes the proof.

Observe that the balls $B_\infty(u_0, r) = \{u \in L^\infty_X : \|u - u_0\| \leq r\}$ of L^∞_X are decomposable. Combining Theorems 2.2 and 2.11 gives the following corollary:

Corollary 2.12. *Under the same hypothesis of Theorem 2.11, if there exists $u_0 \in L^\infty_X$ such that $I_{f_{u_0}}$ is $\sigma(L_\psi, L_{\psi^*})$-inf-compact, then there exists $M > 0$ such that I_{f_M} is inf-compact, where $f_M(\omega, y) = \inf_{\|x-u_0(\omega)\|\leq M} f(\omega, x, y)$.*

Using the same arguments of Corollary 2.3, we deduce th following:

Corollary 2.13. *Under the same hypothesis of Theorem 2.11, if $I_{f_{u_0}}$ is $\sigma(L_\psi, L_{\psi^*})$ -inf-compact for some $u_0 \in L^\infty_X$, then there exists $M > 0$ such that $\bigcup_{\|u-u_0\|\leq M}\{I_{f_u} \leq \lambda\}$ is weakly relatively compact.*

References

1. H. Attouch, *Variational convergence for functions and operators*, Research notes in mathematics, London, 1983.
2. E. J. Balder, *A general approch to lower semicontinuity and lower closure in optimal control theory*, SIAM. J. Control and Optimiz. **22** (1984), 570–598.
3. G. Bottaro and P. Oppezi, *Semicontinuità infériore di un funzionale integrale dependante da funzioni a valori in uno spazio di Banach*, Boll. U.M.I.(5) **17-B** (1980), 1290–1307.
4. A. Bourass, *Comparaison des fonctionnelles intégrales sur les sélections d'une multiapplication mesurable. Conditions de croissances et inclusion des cônes radiaux*, Séminaire d'analyse convexe, Montpellier-Perpignan, exposé **N10** (1982).
5. A. Bourass, B. Ferrahi and O. Kahlaoui, *Semi-continuité inférieure d'une fonctionnelle intégrale*, Bull. of the Greek Math. Soc. **47** (2003), 31–58.
6. A. Bourass and A. Fougèrs, *Conditions de croissance caractéristique de la sci forte-faible d'une fonctionnelle intégrale*, A.V.M.A.C. Perpignan **84-02** (1984).
7. C. Castaing and P. Clauzure, *Semi-continuité des fonctionnelles intégrales*, Séminaire d'analyse convexe, Montpellier-Perpignan, exposé **N15** (1981).
8. C. Castaing and M. Valadier, *Convex analysis and measurable multifunctions*, Lecture Notes in Math., 580, Springer-Verlag, Berlin and New York, 1977.
9. L. Césari, *Closure, lower closure, and semicontinuity theorems in optimal control*, SIAM. J. Control **9** (1971), 287–315.
10. ———, *Lower semicontinuity and lower closure theorem without semi normality condition*, Annal. Mat. Pura ed App. **98** (1974).
11. I. Ekeland and R. Temem, *Analyse convexe et problèmes variationnels*, Dunod-Gauthier-Villars, 1974.
12. A. Fougèrs, *Coercivité des intégrandes convexes normals (I)*, Séminaire d'analyse convexe, Montpellier, exposé **N19** (1976).
13. E. Giner, *Espaces intégraux de type Orlicz. Dualité, compacités, convergence en mesure et Application à l'optimisation*, Doctorat, Perpignan, 1977.
14. A. D. Ioffé, *On lower semicontinuity of integral function I, II*, SIAM. J. Control and Optimiz. **15** (1977), 521–538 et 991–1000.
15. C. Olech, *Existance theory in optimal control problems-the underlying ideas*, in International Conference on Differential Equation (H.A. Antosiewicz, eds.), Academic Press, New York, 1976, pp. 612–629.
16. ———, *Weak lower semicontinuity of integral functionals*, J. Optimiz. Theory and Appl. **19-1** (1976), 135–142.
17. ———, *A characterization of L_1-weak lower semicontinuity of integral functionals*, Bull. Acad. Polon. Sci. **25** (1977), 135–142.
18. R. T. Rochafellar, *Integral functionals, normal integrands, and measurable selections*, Lecture notes in Math., 543, Springer-Verlag, Berlin and New York, 1976.

In: Inequality Theory and Applications, Volume 4
Editors: Yeol Je Cho et al., pp. 33–45
ISBN 1-59454-874-9

ESTIMATION OF RELATIVE ENTROPY USING NOVEL TAYLOR-LIKE REPRESENTATIONS

P. CERONE[1]

ABSTRACT. Sharp bounds are obtained for perturbed generalised Taylor series. The perturbation involves the arithmetic sum of the upper and lower bounds of the $(n+1)^{th}$ derivative. The sharpest bound is in terms of the one norm of the Appell polynomial which constitutes the coefficients of the derivative of the function to be approximated. The results are demonstrated for the estimation of the Kullback-Leibler distance, Shannon entropy and mututal information.

1. Introduction

A number of authors have recently obtained generalisations of the traditional Taylor series expansion of a function $f(x)$ about a point a assuming sufficient differentiability. A Taylor series representation is a fundamental mechanism for estimation in problems arising in many industrial applications. Estimates of bounds on the remainder have also been procured.

In the current article, a novel Taylor-like representation is developed and is used to estimate the Kullback-Leibler distance, Shannon entropy and mututal information. These are of importance in general information theory and, in particular, coding theory.

Before proceeding further with more generalisations, let us introduce some notation. We shall term polynomials of degree k, W_k as Appell type and say $W_k \in \mathcal{A}$ if they satisfy the condition

$$W_k' = \xi_k W_{k-1}(t), \quad W_0(t) = 1, \quad t \in \mathbb{R}. \tag{1.1}$$

These are so named since Appell studied (1.1) with $\xi_k = k$ in 1880 (see [1]). Polynomials satisfying (1.1) with $\xi_k = 1$ have been termed harmonic polynomials in Matić et al. [11], however a simple scaling will demonstrate that these are Appell.

The following results were obtained by Matić et al. [11], where $P_n(t)$ satisfy (1.1) with $\xi_k = 1$.

Theorem 1. *Let $\{P_n\}_{n\in\mathbb{N}}$ be a sequence of polynomials, that satisfy*

$$P_n'(t) = P_{n-1}(t), \quad P_0(t) = 1, \quad t \in \mathbb{R},\ n \in \mathbb{N},\ n \geq 1. \tag{1.2}$$

Received June 23, 2004.

2000 *Mathematics Subject Classification.* Primary 26D15, 41A58.

Key words and phrases. Perturbed generalised Taylor's formula, Appell polynomials, Sharp bounds, Čebyšev-type functional, Kullback-Leibler distance, Entropy, Mutual information.

[1] School of Computer Science and Mathematics, Victoria University of Technology, PO Box 14428, MCMC 8001, VIC, Australia. E-mail: pc@matilda.vu.edu.au (http://rgmia.vu.edu.au/ cerone)

Further, let $I \subset \mathbb{R}$ be a closed interval and $a \in I$. If $f : I \to \mathbb{R}$ is any function such that for some $n \in \mathbb{N}$, $f^{(n)}$ is absolutely continuous, then, for any $x \in I$

$$f(x) = T_n(f; a, x) + R_n(f; a, x), \tag{1.3}$$

where

$$T_n(f; a, x) = f(a) + \sum_{k=1}^{n} (-1)^{k+1} \left[P_k(x) f^{(k)}(x) - P_k(a) f^{(k)}(a) \right], \tag{1.4}$$

$$R_n(f; a, x) = (-1)^n \int_a^x P_n(t) f^{(n+1)}(t) dt. \tag{1.5}$$

They also pointed out the following bounds for the remainder $R_n(f; a, x)$.

Corollary 1. *With the above assumptions of Theorem 1, the following estimations hold. Namely, for $x \geq a$,*

$$\begin{aligned} &|R_n(f; a, x)| \\ &\leq \begin{cases} \|P_n\|_\infty \left\|f^{(n+1)}\right\|_1 & \text{provided } f^{(n+1)} \in L_1[a, x], \\ \|P_n\|_q \left\|f^{(n+1)}\right\|_p & \text{provided } f^{(n+1)} \in L_p[a, x],\ p > 1,\ \frac{1}{p} + \frac{1}{q} = 1, \\ \|P_n\|_1 \left\|f^{(n+1)}\right\|_\infty & \text{provided } f^{(n+1)} \in L_\infty[a, x], \end{cases} \end{aligned} \tag{1.6}$$

where $x \geq a$ and $\|\cdot\|_s$ $(1 \leq s \leq \infty)$ are the usual s–Lebesgue norms. That is,

$$\|g\|_s := \left(\int_a^x |g(t)|^s \, dt \right)^{\frac{1}{s}}, \quad s \in [1, \infty),$$

and

$$\|g\|_\infty := \operatorname*{ess\,sup}_{t \in [a,x]} |g(t)|.$$

Let

$$P_n^{c\lambda}(t) = \frac{(t - \theta(\lambda))^n}{n!}, \quad \theta(\lambda) = \lambda a + (1 - \lambda) x, \quad \lambda \in [0, 1], \tag{1.7}$$

represent polynomials involving; a convex combination of the end points. Although the methodology is applicable for general Appell type polynomials such as Euler and Bernoulli, only the above set will be considered in detail here. We should note that the dependence of the polynomials in (1.7) on x is not shown explicitly.

With the polynomials (1.7), then, from (1.4) and (1.5),

$$\begin{aligned} &T_n^{c\lambda}(f; a, x) \\ &= f(a) + \sum_{k=1}^{n} (-1)^{k+1} \frac{(x-a)^k}{k!} \left[\lambda^k f^{(k)}(x) + (-1)^{k+1} (1-\lambda)^k f^{(k)}(a) \right] \end{aligned} \tag{1.8}$$

and the remainder

$$\begin{aligned} R_n^{c\lambda}(f; a, x) &= \frac{(-1)^{n+1}}{n!} \int_a^x (t - \theta(\lambda))^n f^{(n+1)}(t) dt, \\ \theta(\lambda) &= \lambda a + (1 - \lambda) x, \quad \lambda \in [0, 1]. \end{aligned} \tag{1.9}$$

The above expression (1.8)–(1.9) was obtained by Matić et al. [11] only for the equivalent of $\lambda = 0$ and $\frac{1}{2}$ in (1.7).

Cerone and Dragomir [4] obtained the following theorem which follows directly from Corollary 1 with $P_n^{c\lambda}(t)$ as given by (1.7) and, using (1.8) and (1.9).

Theorem 2. *Assume that f is as in Theorem 1 with $x \geq a$, then we have*

$$(1.10)\qquad \begin{aligned} &\left|f(x) - T_n^{c_\lambda}(f;a,x)\right| = \left|R_n^{c_\lambda}(f;a,x)\right| \\ &\leq \begin{cases} \frac{1}{n!}(x-a)^n\left[\frac{1}{2}+\left|\lambda-\frac{1}{2}\right|\right]^n \left\|f^{(n+1)}\right\|_1 & \text{if } f^{(n+1)} \in L_1[a,x]; \\ \frac{1}{n!(nq+1)^{\frac{1}{q}}}(x-a)^{n+\frac{1}{q}}\left[(1-\lambda)^{nq+1}+\lambda^{nq+1}\right]^{\frac{1}{q}}\left\|f^{(n+1)}\right\|_p & \\ \qquad \text{if } f^{(n+1)} \in L_p[a,x],\ p>1,\ \frac{1}{p}+\frac{1}{q}=1; & \\ \frac{1}{(n+1)!}(x-a)^{n+1}\left[(1-\lambda)^{n+1}+\lambda^{n+1}\right]\left\|f^{(n+1)}\right\|_\infty, & \\ \qquad \text{if } f^{(n+1)} \in L_\infty[a,x]. & \end{cases} \end{aligned}$$

It was also noted that since

$$h_1(\lambda) = \left[\frac{1}{2} + \left|\lambda - \frac{1}{2}\right|\right]^n, \quad h_2(\lambda) = \left[(1-\lambda)^{nq+1} + \lambda^{nq+1}\right]^{\frac{1}{q}}$$

and

$$h_3(\lambda) = (1-\lambda)^{n+1} + \lambda^{n+1}$$

are convex and symmetric about $\frac{1}{2}$, then

$$\inf_{\lambda \in [0,1]} h_i(\lambda) = h_i\left(\frac{1}{2}\right), \qquad i = 1,2,3.$$

Hence the best inequality possible in the class, in the sense of providing the tightest bound is obtained from taking $\lambda = \frac{1}{2}$.

Taking $\lambda = 0$ in (1.10) produces the classical Taylor series expansion in terms of the $L_p[a,x]$, $p \geq 1$, Lebesgue norms for the bounds (see, for example, Dragomir [8]).

Recently, in [7], Dragomir introduced a perturbed Taylor's formula using the Grüss inequality for the Čebyšev functional. Matić et al. [11] obtained generalised Taylor's formulae involving expansions in terms of general polynomials satisfying (1.2) producing in particular Theorem 1 and Corollary 1 above. They also examined perturbed versions of (1.3), namely

$$(1.11)\qquad f(x) = T_n(f;a,x) + (-1)^n\left[P_{n+1}(x) - P_{n+1}(a)\right]\left[f^{(n)};a,x\right] + \rho_n(f;a,x),$$

where

$$(1.12)\qquad \left[f^{(n)};a,x\right] := \frac{f^{(n)}(x) - f^{(n)}(a)}{x-a}, \text{ the divided difference,}$$

$$(1.13)\qquad \rho_n(f;a,x) \text{ is the remainder.}$$

Dragomir [7] developed an estimate of the remainder using the Grüss inequality for $P_n(t) = \frac{(t-x)^n}{n!}$, Matić et al. [11] used a premature or pre-Grüss argument to procure bounds on $\rho_n^{c_{\frac{1}{2}}}(f;a,x)$, $\rho_n^{c_0}(f;a,x)$. Dragomir [8] obtained tighter bounds for the same polynomial generators of the perturbed Taylor series for $f^{(n+1)} \in L_2(I)$ with $x,a \in I \subseteq \mathbb{R}$. In the paper [4], Cerone and Dragomir procured bounds on $\rho_n(f;a,x)$ in terms of $\Delta-$seminorms resulting from the Čebyšev functional and Korkine's identity.

In [7], S.S. Dragomir seems to be the first author to introduce the perturbed Taylor formula

$$(1.14)\qquad f(x) = T_n(f;a,x) + \frac{(x-a)^{n+1}}{(n+1)!}\left[f^{(n)};a,x\right] + \rho_n(f;a,x),$$

where

$$T_n(f;a,x) = \sum_{k=0}^{n} \frac{(x-a)^k}{k!} f^{(k)}(a) \tag{1.15}$$

and $[f^{(n)};a,x]$ is as given in (1.12). Dragomir [7] estimated the remainder $\rho_n(f;a,x)$ by using Grüss and Čebyšev type inequalities.

In [11], the authors generalised and improved the results from [D1] via a pre-Grüss inequality (see [11, Theorem 3]).

Theorem 3. *Let $\{P_n\}_{n\in\mathbb{N}}$ be a sequence of polynomials satisfying (1.2). Let $I \subset \mathbb{R}$ be a closed interval and $a \in I$. Suppose $f : I \to \mathbb{R}$ is as in Theorem 1. Then for all $x \in I$ we have the perturbed generalised Taylor formula (1.11), where $x \geq a$, the remainder $\rho(f;a,x)$ satisfies the estimate*

$$|\rho_n(f;a,x)| \leq \frac{x-a}{2}\sqrt{\mathcal{T}(P_n,P_n)}\,[\Gamma(x)-\gamma(x)], \tag{1.16}$$

provided that $f^{(n+1)}$ is bounded and

$$\gamma(x) := \inf_{t\in[a,x]} f^{(n+1)}(t) > -\infty, \quad \Gamma(x) := \sup_{t\in[a,x]} f^{(n+1)}(t) < \infty. \tag{1.17}$$

In (1.16), $\mathcal{T}(\cdot,\cdot)$ is the Čebyšev functional on the interval $[a,x]$. That is,

$$\mathcal{T}(g,h) := \frac{1}{x-a}\int_a^x g(t)h(t)dt - \frac{1}{x-a}\int_a^x g(t)dt \cdot \frac{1}{x-a}\int_a^x h(t)dt. \tag{1.18}$$

It is the intention in the current article to produce perturbed generalised Taylor series like (1.11), however the perturbation involves the arithmetic average of the upper and lower bounds of the $f^{(n+1)}(t)$, $t \in I$. The bounds for the expansion involve the norms of the Appell polynomials with the one norm, which is shown to provide the tightest bound.

A novel Čebyšev functional and its bounds are presented in Section 2, the results of which, are applied to perturbed generalised Taylor series. The approximation of the logarithmic function using the results of Section 2 is presented in Section 3. The results are demonstrated by estimating some key entities in information and coding theory. In particular, the Kullback-Leibler distance, Shannon's entropy and mutual information are approximated by rational functions providing sharp bounds. This application in Information Theory will be discussed in detail in Section 4.

2. The Čebyšev-like Functional and Perturbed Taylor Results with Bounds

Let $f,g : [a,b] \to \mathbb{R}$ be two integrable functions and define the functional

$$T(f,g;a,b) := \mathcal{M}(fg;a,b) - \mathcal{M}(f;a,b)\,\mathcal{M}(g;a,b), \tag{2.1}$$

where the integral mean is given by

$$\mathcal{M}(f;a,b) := \frac{1}{b-a}\int_a^b f(x)dx. \tag{2.2}$$

The functional $T(f,g;a,b)$ as defined in (2.1)–(2.2) is widely known in the literature as Čebyšev's functional. The reader is referred to [12], Chapters IX and X, and to the work by Dragomir [6] and Fink [10] for extensive treatments of the functional.

We now define a Čebyšev-like functional

$$C(f,g;a,b) := \mathcal{M}(fg;a,b) - \frac{M+m}{2}\mathcal{M}(f;a,b), \tag{2.3}$$

where $-\infty < m \le g(t) \le M < \infty$ for $t \in [a,b]$ and $\mathcal{M}(f;a,b)$ is as given by (2.2).

The following theorem holds providing bounds for the functional $C(f,g;a,b)$ (see Cerone [3]). The proof involves using the identity

$$\mathcal{M}(fg;a,b) - \frac{M+m}{2}\mathcal{M}(f;a,b) = \mathcal{M}\left(f\left(g(\cdot) - \frac{M+m}{2}\right)\right). \tag{2.4}$$

Theorem 4. *Let $f,g:[a,b]\to\mathbb{R}$ be integrable functions and $-\infty < m \le g(t) \le M < \infty$, then*

$$\begin{aligned}
&|C(f,g;a,b)| \\
&= \left|\mathcal{M}(fg;a,b) - \frac{M+m}{2}\mathcal{M}(f;a,b)\right| \\
&\le \frac{M-m}{2}\cdot\frac{1}{b-a}\|f\|_1, \quad f\in L_1[a,b], \\
&\le \frac{M-m}{2}\cdot\frac{1}{(b-a)^{\frac{1}{p}}}\|f\|_p, \quad f\in L_p[a,b],\ 1<p<\infty, \\
&\le \frac{M-m}{2}\|f\|_\infty = \frac{M-m}{2}\max\{|N|,|n|\}, \quad f\in L_\infty[a,b], \\
&\qquad -\infty < n \le f(t) \le N < \infty,\ t\in[a,b],
\end{aligned} \tag{2.5}$$

where $\|f\|_p$ are the usual Lebesgue norms for $f\in L_p[a,b]$ defined by

$$\|f\|_p := \left(\int_a^b |f(t)|^p\,dt\right)^{\frac{1}{p}}, \quad 1\le p<\infty,$$

and

$$\|f\|_\infty := \operatorname*{ess\,sup}_{t\in[a,b]}|f(t)|.$$

The $\frac{1}{2}$ in the three inequalities in (2.5) are sharp.

Remark 1. The inequalities in (2.5) are in the order of increasing coarseness although each of them are sharp for $f\in L_p[a,b]$, $p\ge 1$.

We now apply the above results to obtain sharp bounds for perturbed Taylor-like formulae.

Theorem 5. *Let $\{P_n\}_{n\in\mathbb{N}}$ be a sequence of polynomials that satisfy (1.2). Further, let $I\subset\mathbb{R}$ be a closed interval and $a\in I$. If $f:I\to\mathbb{R}$ is a function which for some $n\in\mathbb{N}$, $f^{(n)}$ is absolutely continuous and $-\infty<\gamma_{n+1}(x)\le f^{(n+1)}(t)\le\Gamma_{n+1}(x)<\infty$, then, for any $a\le x\in I$,*

$$\begin{aligned}
f(x) &= T_n(f;a,x) + \left(\frac{\gamma_{n+1}(x)+\Gamma_{n+1}(x)}{2}\right)(-1)^n\frac{[P_{n+1}(x)-P_{n+1}(a)]}{n+1} \\
&\quad + G_n(f;a,x)
\end{aligned} \tag{2.6}$$

and

$$\begin{aligned}
& |G_n(f;a,x)| \\
& \le \frac{\Gamma_{n+1}(x)-\gamma_{n+1}(x)}{2}\|P_n\|_{1,[a,x]}, \quad P_n \in L_1[I], \\
(2.7)\qquad & \le \frac{\Gamma_{n+1}(x)-\gamma_{n+1}(x)}{2}\cdot\frac{1}{(x-a)^{\frac{1}{p}-1}}\|P_n\|_{p,[a,x]}, \ P_n \in L_p[I], 1<p<\infty, \\
& \le \frac{\Gamma_{n+1}(x)-\gamma_{n+1}(x)}{2}\|P_n\|_{\infty,[a,x]}, \quad P_n \in L_\infty[I],
\end{aligned}$$

where

$$(2.8)\qquad G_n(f;a,x) = (-1)^n \int_a^x P_n(t)\left[f^{(n+1)}(t) - \frac{\gamma_{n+1}(x)+\Gamma_{n+1}(x)}{2}\right]dt$$

and $T_n(f;a,x)$ is as given by (1.4).

Proof. If we identify $(-1)^n P_n(t)$ with $f(t)$ and $f^{(n+1)}(t)$ with $g(t)$ in Theorem 4, then, from identity (2.4), we have

$$\begin{aligned}
G_n(f;a,x) &= (-1)^n \int_a^x P_n(t)\left[f^{(n+1)}(t) - \frac{\gamma_{n+1}(x)+\Gamma_{n+1}(x)}{2}\right]dt \\
(2.9)\qquad &= (x-a)C\left((-1)^n P_n(t), f^{(n+1)}(t);a,x\right).
\end{aligned}$$

That is,

$$\begin{aligned}
& G_n(f;a,x) \\
(2.10)\qquad & = (-1)^n \int_a^x P_n(t)f^{(n+1)}(t)dt - (-1)^n \frac{\gamma_{n+1}(x)+\Gamma_{n+1}(x)}{2}\int_a^x P_n(t)dt \\
& = R_n(f;a,x) - (-1)^n\left(\frac{\gamma_{n+1}(x)+\Gamma_{n+1}(x)}{2}\right)\frac{[P_{n+1}(x)-P_{n+1}(a)]}{n+1},
\end{aligned}$$

where $R_n(f;a,x)$ satisfies (1.3)–(1.5). Using identity (1.3) in (2.10) produces the stated result (2.6).

For the bound on the remainder $|G_n(f;a,x)|$, we have from the first inequality in (2.5) and (2.9)

$$|G_n(f;a,x)| \le \frac{\Gamma_{n+1}(x)-\gamma_{n+1}(x)}{2}\int_a^x |P_n(t)|\,dt$$

and hence (2.7) follows. Utilising the second and third inequality in (2.5) produces the respective bounds in (2.7).

Corollary 2. *Let the conditions of Theorem 5 persist and assume $a \le x$, then we have, from (2.6) and (2.7), for $P_n^{c\lambda}(t)$ given by (1.7)*

$$\begin{aligned}
& \Bigg| f(x) - T_n^{c\lambda}(f;a,x) - \frac{1}{(n+1)!}\left[(\theta(\lambda)-a)^{n+1}\right. \\
& \qquad \left. + (-1)^n (x-\theta(\lambda))^{n+1}\right]\left(\frac{\gamma_{n+1}(x)+\Gamma_{n+1}(x)}{2}\right)\Bigg| \\
(2.11)\qquad & \le \frac{\Gamma_{n+1}(x)-\gamma_{n+1}(x)}{2}\cdot\frac{1}{n!}\Psi_1(\lambda;a,x) \\
& \le \frac{\Gamma_{n+1}(x)-\gamma_{n+1}(x)}{2}\cdot\frac{1}{(x-a)^{\frac{1}{p}-1}}\cdot\frac{1}{n!}\left[\Psi_p(\lambda;a,x)\right]^{\frac{1}{p}} \\
& \le \frac{\Gamma_{n+1}(x)-\gamma_{n+1}(x)}{2}\cdot\frac{(x-a)}{n!}\cdot\left[\frac{x-a}{2}+\left|\theta(\lambda)-\frac{a+x}{2}\right|\right]^n,
\end{aligned}$$

where

$$\text{(2.12)}\qquad \begin{cases} \Psi_p(\lambda; a, x) = \dfrac{(x-\theta(\lambda))^{np+1} + (\theta(\lambda)-a)^{np+1}}{np+1} \\ \theta(\lambda) = \lambda a + (1-\lambda)x, \quad \lambda \in [0,1]. \end{cases}$$

Proof. We need to evaluate for $P_n^{c_\lambda}(t)$ given by (1.7)

$$\int_a^x P_n^{c_\lambda}(t)dt \text{ and } \int_a^x |P_n^{c_\lambda}(t)|^p dt, \quad p \geq 1.$$

Thus, from (1.7), we have

$$\int_a^x P_n^{c_\lambda}(t)dt = \frac{1}{(n+1)!}\left[(x-\theta(\lambda))^{n+1} - (\theta(\lambda)-a)^{n+1}\right]$$

and

$$\begin{aligned} \int_a^x |P_n^{c_\lambda}(t)|^p dt &= \frac{1}{n!}\int_a^x |t-\theta(\lambda)|^n dt \\ &= \frac{1}{n!}\left[\int_a^{\theta(\lambda)} (\theta(\lambda)-t)^n dt + \int_{\theta(\lambda)}^b (t-\theta(\lambda))^n dt\right] \\ &= \frac{1}{(n+1)!}\left[(\theta(\lambda)-a)^{n+1} + (x-\theta(\lambda))^{n+1}\right], \end{aligned}$$

producing (2.11). Further,

$$\begin{aligned} \|P_n^{c_\lambda}\|_{p,[a,x]}^p &= \int_a^x |P_n^{c_\lambda}(t)|^p dt = \frac{1}{n!}\int_a^x |t-\theta(\lambda)|^{np} dt \\ &= \frac{1}{n!}\left[\int_a^{\theta(\lambda)} (\theta(\lambda)-t)^{np} dt + \int_{\theta(\lambda)}^x (t-\theta(\lambda))^{np} dt\right] \\ &= \frac{1}{n!}\left[\frac{(\theta(\lambda)-a)^{np+1} + (x-\theta(\lambda))^{np+1}}{np+1}\right] \end{aligned}$$

and so, from (2.7), the second inequality in (2.11) is procured.

The final inequality is obtained from (1.7) and (2.7) giving

$$\begin{aligned} \|P_n^{c_\lambda}\|_{\infty,[a,x]} &= \operatorname*{ess\,sup}_{t\in[a,x]} |P_n^{c_\lambda}(t)| \\ &= \operatorname*{ess\,sup}_{t\in[a,x]} \frac{|t-\theta(\lambda)|^n}{n!} \\ &= \frac{1}{n!}\left[\max\{x-\theta(\lambda), \theta(\lambda)-a\}\right]^n. \end{aligned}$$

This completes the proof.

Remark 2. The bounds in (2.11) are in order of increasing coarseness. This was commented upon in Remark 1 referring to the results (2.5), of which (2.11) is a specialisation. For $\lambda = \frac{1}{2}$, $\theta\left(\frac{1}{2}\right) = \frac{a+x}{2}$, then, from (2.11) and (2.12),

$$\begin{aligned}
&\left| f(x) - T_n^{c\frac{1}{2}}(f;a,x) - \frac{1+(-1)^n}{(n+1)!} \cdot \left(\frac{x-a}{2}\right)^{n+1} \left(\frac{\gamma_{n+1}(x) + \Gamma_{n+1}(x)}{2}\right) \right| \\
&\le \frac{\Gamma_{n+1}(x) - \gamma_{n+1}(x)}{2} \cdot \frac{(x-a)^{n+1}}{2^n (n+1)!} \\
&\le \frac{\Gamma_{n+1}(x) - \gamma_{n+1}(x)}{2} \cdot \frac{(x-a)^{n+1}}{2^n (np+1)^{\frac{1}{p}} n!} \\
&\le \frac{\Gamma_{n+1}(x) - \gamma_{n+1}(x)}{2} \cdot \frac{(x-a)^{n+1}}{2^n n!},
\end{aligned} \tag{2.13}$$

where, from (1.8),

$$T_n^{c\frac{1}{2}}(f;a,x) = f(a) + \sum_{k=1}^{n} \frac{(x-a)^k}{2^k k!} \left[f^{(k)}(a) + (-1)^{k+1} f^{(k)}(x) \right], \tag{2.14}$$

from which we may confirm, for this case at least, the fact that the bounds are in order of increasing coarseness since $\frac{1}{n+1} < \frac{1}{(np+1)^{\frac{1}{p}}} < 1$, $1 < p < \infty$.

We further note that for n odd, the perturbation in (2.13) vanishes, giving the tightest bound

$$\left| f(x) - T_n^{c\frac{1}{2}}(f;a,x) \right| \le \frac{\Gamma_{n+1}(x) - \gamma_{n+1}(x)}{2} \cdot \frac{(x-a)^{n+1}}{2^n (n+1)!}, \quad n \text{ odd}. \tag{2.15}$$

For n odd, the result (2.15) may be compared with the last result in (1.10) with $\lambda = \frac{1}{2}$, demonstrating that it is tighter since

$$\|f^{(n+1)}\|_\infty \ge \frac{\Gamma_{n+1}(x) - \gamma_{n+1}(x)}{2},$$

where $\gamma_{n+1}(x) \le f(t) \le \Gamma_{n+1}(x)$, $t \in [a,x]$.

For n even, the bound is still tighter in (2.13), however, the perturbation is now present.

If $\lambda = 0$ in (2.11), then $\theta(0) = x$ and we obtain a perturbed version of the traditional Taylor series expansion about a point a. That is, from (1.8),

$$\begin{aligned}
&\left| f(x) - T_n^{c0}(f;a,x) - \frac{(x-a)^{n+1}}{(n+1)!} \left(\frac{\gamma_{n+1}(x) + \Gamma_{n+1}(x)}{2}\right) \right| \\
&= \left| f(x) - \sum_{k=0}^{n} \frac{(x-a)^k}{k!} f^{(k)}(a) - \frac{(x-a)^{n+1}}{(n+1)!} \left(\frac{\gamma_{n+1}(x) + \Gamma_{n+1}(x)}{2}\right) \right| \\
&\le \frac{\Gamma_{n+1}(x) - \gamma_{n+1}(x)}{2} \cdot \frac{(x-a)^{n+1}}{(n+1)!},
\end{aligned} \tag{2.16}$$

where

$$T_n^{c0}(f;a,x) = f(x) + \sum_{k=0}^{n} (-1)^{k+1} \frac{(x-a)^k}{k!} f^{(k)}(a).$$

We notice that the bound in (2.16) is inferior to that in the first inequality in (2.13) where $T_n^{c_{\frac{1}{2}}}(f;a,x)$ is as given by (2.14). However, (2.14) requires information involving $f^{(k)}(x)$ being available in order to approximate $f(x)$.

3. Applications to the Logarithm

We shall apply the results of the previous sections to the logarithm function to illustrate the results.

Let $f:(0,\infty)\to\mathbb{R}$, $f(t)=\ln t$, then

$$f^{(k)}(t)=(-1)^{k-1}\frac{(k-1)!}{t^k},\quad t>0,\ k\in\mathbb{N}. \tag{3.1}$$

We note that $f^{(n+1)}$ is strictly monotonic on $(0,\infty)$ so that

$$\begin{aligned}&\frac{\gamma_{n+1}(x)\mp\Gamma_{n+1}(x)}{2}\\&=\frac{1}{2}\left[\max\left\{f^{(n+1)}(a),f^{(n+1)}(x)\right\}\mp\min\left\{f^{(n+1)}(a),f^{(n+1)}(x)\right\}\right]\\&=\frac{1}{2}\left|f^{(n+1)}(a)\mp f^{(n+1)}(x)\right|\\&=\frac{n!}{2}\left(\frac{1}{a^{n+1}}\mp\frac{1}{x^{n+1}}\right),\quad x\geq a.\end{aligned} \tag{3.2}$$

We shall utilise the first inequality in (2.11), which is the least coarse of the three, to illustrate the results to give

$$\begin{aligned}&\left|f(x)-T_n^{c_\lambda}(\ln;a,x)-\frac{1}{2(n+1)}\left[(\theta(\lambda)-a)^{n+1}+(-1)^n(x-\theta(\lambda))^{n+1}\right]\right|\\&\leq\frac{1}{2(n+1)}\left(\frac{1}{a^{n+1}}-\frac{1}{x^{n+1}}\right)\left[(\theta(\lambda)-a)^{n+1}+(x-\theta(\lambda))^{n+1}\right],\end{aligned} \tag{3.3}$$

where $\theta(\lambda)=\lambda a+(1-\lambda)x$, $\lambda\in[0,1]$, and, from (1.8),

$$T_n^{c_\lambda}(\ln;a,x)=\ln a+\sum_{k=1}^{n}\frac{(x-a)^k}{k}\left[\left(\frac{\lambda}{x}\right)^k+(-1)^{k+1}\left(\frac{1-\lambda}{a}\right)^k\right].$$

Simplification of (3.1) gives

$$\begin{aligned}&\left|\ln x-\ln a-\sum_{k=1}^{n}\frac{(x-a)^k}{k}\left[\left(\frac{\lambda}{x}\right)^k+(-1)^{k+1}\left(\frac{1-\lambda}{a}\right)^k\right]\right.\\&\qquad\left.-\frac{(x-a)^{n+1}}{2(n+1)}\left[(1-\lambda)^{n+1}+\lambda^{n+1}\right]\left(\frac{1}{a^{n+1}}+\frac{1}{x^{n+1}}\right)\right|\\&\leq\frac{(x-a)^{n+1}}{2(n+1)}\left[(1-\lambda)^{n+1}+\lambda^{n+1}\right]\left(\frac{1}{a^{n+1}}-\frac{1}{x^{n+1}}\right),\end{aligned} \tag{3.4}$$

where the sharpest bound results from taking $\lambda=\frac{1}{2}$.

In particular, taking $\lambda=0$ gives

$$\begin{aligned}&\left|\ln x-\ln a-\sum_{k=1}^{n}\frac{(-1)^{k+1}}{k}\left(\frac{x-a}{a}\right)^k-\frac{(x-a)^{n+1}}{2(n+1)}\left(\frac{1}{a^{n+1}}+\frac{1}{x^{n+1}}\right)\right|\\&\leq\frac{(x-a)^{n+1}}{2(n+1)}\left(\frac{1}{a^{n+1}}-\frac{1}{x^{n+1}}\right):=R,\end{aligned} \tag{3.5}$$

which may be compared with the result (see case in Section 4 of Matić et al. [11])

$$\ln x = \ln a + \sum_{k=1}^{n} \frac{(-1)^{k+1}}{k}\left(\frac{x-a}{a}\right)^k + (-1)^n \frac{(x-a)^n}{n(n+1)}\left(\frac{1}{a^n} - \frac{1}{x^n}\right) + B, \tag{3.6}$$

where $|B| \leq \frac{n}{\sqrt{2n+1}} R$.

It may be seen that the bound R obtained here, is much tighter, especially for large n since $\frac{n}{\sqrt{2n+1}} > 1$. It must be remembered, however, that a different perturbation is present in (3.5) than in (3.6). There is very little difference in complexity between the two perturbations. The perturbation used in [11] from the traditional Čebyšev functional (2.1) giving rise to (1.11) rather than the novel Čebyšev functional (2.3) producing (2.6). That is, the perturbation in (1.11) involves $\left[f^{(n+1)}; a, x\right] = \frac{f^{(n)}(x) - f^{(n)}(a)}{x-a}$ whereas that in (2.6) contains $\frac{\gamma_{n+1}(x) + \Gamma_{n+1}(x)}{2}$, where $\gamma_{n+1}(x) \leq f^{(n+1)}(t) \leq \Gamma_{n+1}(x)$.

4. Approximation of the Kullback-Leibler Distance and Related Measures

The results of Section 3 will now be applied to approximate the *Kullback-Leibler distance* or *relative entropy* $D(p||q)$ where for probability mass functions $p(x)$, $q(x)$

$$D(p||q) := \sum_{x\in\chi} p(x) \ln\left(\frac{p(x)}{q(x)}\right), \tag{4.1}$$

with $0 \ln\left(\frac{0}{q}\right) = 0$ and $p \ln\left(\frac{p}{0}\right) = \infty$,

The relative entropy $D(p||q)$ is a measure of inefficiency in assuming a distribution $q(\cdot)$ when the true distribution is $p(\cdot)$. In coding theory, for example, if the true distribution of a random variable X with mass function $p(x)$, $x \in \chi$, were known, then a code could be constructed with average description length given by the *Shannon entropy*

$$H(p) := -\sum_{x\in\chi} p(x) \ln p(x). \tag{4.2}$$

If the code for a mass function $q(x)$ were used instead, then on average $H(p) + D(p||q)$ bits would be required to describe the random variable [5, p. 18]. The *mutual information*, $I(X,Y)$ is a measure of the amount of information contained by one random variable about the other. It represents the effect in reducing uncertainty of one random variable from knowledge of the other. *Mutual information* is defined by

$$I(X,Y) := D(\rho(x,y)||p(x)q(y)), \tag{4.3}$$

where $\rho(x,y)$ is the joint probability mass function and $p(x)$, $q(y)$ are its marginals.

Theorem 6. *Let X and Y be random variables with probability mass functions $p(x)$ and $q(x)$ respectively with $x \in \chi$. Then we have the following approximation for $D(p||q)$. Namely,*

$$\begin{aligned} &\left| D(p||q) - \sum_{x\in\chi} p(x) \sum_{k=1}^{n} \frac{(p(x)-q(x))^k}{k}\left[\left(\frac{\lambda}{p(x)}\right)^k + (-1)^{k+1}\left(\frac{1-\lambda}{q(x)}\right)^k\right]\right. \\ &\left. - \frac{(1-\lambda)^{n+1} + \lambda^{n+1}}{2(n+1)} \sum_{x\in\chi} p(x)(p(x)-q(x))^{n+1}\left(\frac{1}{q^{n+1}(x)} + \frac{1}{p^{n+1}(x)}\right)\right| \\ &\leq \frac{(1-\lambda)^{n+1} + \lambda^{n+1}}{2(n+1)} \sum_{x\in\chi} p(x)|p(x)-q(x)|^{n+1}\left|\frac{1}{q^{n+1}(x)} - \frac{1}{p^{n+1}(x)}\right|. \end{aligned} \tag{4.4}$$

Proof. From (3.4), we have, for all $a, b > 0$,

$$\begin{aligned}(4.5)\quad & \left| \ln b - \ln a - \sum_{k=1}^{n} \frac{(b-a)^k}{k} \left[\left(\frac{\lambda}{b}\right)^k + (-1)^{k+1} \left(\frac{1-\lambda}{a}\right)^k \right] \right. \\ & \left. - \frac{(1-\lambda)^{n+1} + \lambda^{n+1}}{2(n+1)} (b-a)^{n+1} \left(\frac{1}{a^{n+1}} + \frac{1}{b^{n+1}} \right) \right| \\ & \leq \frac{(1-\lambda)^{n+1} + \lambda^{n+1}}{2(n+1)} |b-a|^{n+1} \left| \frac{1}{a^{n+1}} - \frac{1}{b^{n+1}} \right|,\end{aligned}$$

where the modulus in the bound has been included to accommodate the lack of knowledge about the relative position of a and b.

Choose $a = q(x)$ and $b = p(x)$ to give, after multiplication by $p(x)$ and summation over $x \in \chi$, the stated expression (4.4) follows.

Corollary 3. *Let the conditions of Theorem 6 hold, then*

$$\begin{aligned}(4.6)\quad & \left| D(p||q) - \sum_{x\in\chi} p(x) \sum_{k=1}^{n} \frac{(p(x)-q(x))^k}{k \cdot 2^k} \left[\frac{1}{p^k(x)} + \frac{(-1)^{k+1}}{q^k(x)} \right] \right. \\ & \left. - \frac{1}{(n+1)2^{n+1}} \sum_{x\in\chi} p(x) (p(x)-q(x))^{n+1} \left(\frac{1}{q^{n+1}(x)} + \frac{1}{p^{n+1}(x)} \right) \right| \\ & \leq \frac{1}{(n+1)2^{n+1}} \sum_{x\in\chi} p(x) |p(x)-q(x)|^{n+1} \left| \frac{1}{q^{n+1}(x)} - \frac{1}{p^{n+1}(x)} \right|.\end{aligned}$$

Proof. Taking $\lambda = \frac{1}{2}$ provides the tightest bound in (4.4) giving the result (4.6).

Corollary 4. *For the conditions of Theorem 6 continuing to hold then*

$$(4.7)\quad 0 \leq D(p||q) \leq \sum_{x\in\chi} \frac{p^2(x)}{q(x)} - 1,$$

$$\begin{aligned}(4.8)\quad & \left| D(p||q) - (1-\lambda) \left[\sum_{x\in\chi} p(x) \left(\frac{p(x)}{q(x)} \right) - 1 \right] \right. \\ & \left. - \frac{1}{2} \left[\frac{1}{4} + \left(\lambda - \frac{1}{2} \right)^2 \right] \sum_{x\in\chi} p(x) (p(x) - q(x))^2 \left[\frac{1}{q^2(x)} + \frac{1}{p^2(x)} \right] \right| \\ & \leq \frac{1}{2} \left[\frac{1}{4} + \left(\lambda - \frac{1}{2} \right)^2 \right] \sum_{x\in\chi} p(x) (p(x) - q(x))^2 \left| \frac{1}{q^2(x)} - \frac{1}{p^2(x)} \right|.\end{aligned}$$

Proof. Taking $n = 0$ in (4.4) recaptures the result (4.7) of [9] and [5]. The result (4.8) follows after some simplification for (4.4) on taking $n = 1$.

Remark 3. Equation (4.8) may be rearranged to give upper and lower bounds for $D(p||q)$ in terms of rationals functions of $p(x)$ and $q(x)$.

The following theorem provides approximations for the Shannon entropy $H(p)$, as defined by (4.2) together with bounds.

Theorem 7. *Let X be a random variable with probability mass function $p(x)$. Then, for $x \in \chi$,*

$$\left| H(p) - \ln|\chi| + \sum_{x\in\chi}\sum_{k=1}^{n} \frac{(|\chi|\,p(x)-1)^k}{k\,|\chi|^k\,p^{k-1}(x)}\left[\lambda^k + (-1)^{k+1}|\chi|^k(1-\lambda)^k\right] \right.$$

$$\left. + \frac{(1-\lambda)^{n+1}+\lambda^{n+1}}{2(n+1)} \sum_{x\in\chi} \frac{(|\chi|\,p(x)-1)^{n+1}}{|\chi|^{2(n+1)}\,p^n(x)}\left(p^{n+1}(x)+|\chi|^{n+1}\right) \right| \tag{4.9}$$

$$\leq \frac{(1-\lambda)^{n+1}+\lambda^{n+1}}{2(n+1)} \sum_{x\in\chi} \frac{\left||\chi|\,p(x)-1\right|^{n+1}}{|\chi|^{2(n+1)}\,p^n(x)}\left|p^{n+1}(x)+|\chi|^{n+1}\right|.$$

Proof. From (4.1) and (4.2), we note that

$$H(p) = \ln|\chi| - D(p||u),$$

where $u(x) = \frac{1}{|\chi|}$, the uniform probability mass function on χ. Thus, from (4.4), taking $q(x) = u(x) = \frac{1}{|\chi|}$ gives (4.9) upon some simplification.

The following theorem gives an approximation to the *mutual information* as defined by (4.3) with (4.1). Namely, for random variables X and Y with joint probability mass function $\rho(x,y)$ and marginals $p(x)$, $q(y)$, $x \in \mathcal{X}$, $y \in \mathcal{Y}$,

$$I(X,Y) = \sum_{x\in\mathcal{X}}\sum_{y\in\mathcal{Y}} \rho(x,y)\ln\frac{\rho(x,y)}{p(x)q(y)}. \tag{4.10}$$

Theorem 8. *With the above assumptions regarding X and Y, we have*

$$\left| I(X,Y) - \sum_{x\in\mathcal{X}}\sum_{y\in\mathcal{Y}} \rho(x,y) \sum_{k=1}^{n} \rho(x,y)\frac{(\rho(x,y)-p(x)q(y))^k}{k} \right.$$

$$\times\left[\left(\frac{\lambda}{\rho(x,y)}\right)^k + (-1)^{k+1}\left(\frac{1-\lambda}{p(x)q(y)}\right)^k\right] - \frac{(1-\lambda)^{n+1}+\lambda^{n+1}}{2(n+1)}$$

$$\left. \times \sum_{x\in\mathcal{X}}\sum_{y\in\mathcal{Y}} \rho(x,y)(\rho(x,y)-p(x)q(y))^{n+1}\left(\frac{1}{p^n(x)q^n(x)}+\frac{1}{\rho^n(x,y)}\right) \right| \tag{4.11}$$

$$\leq \frac{(1-\lambda)^{n+1}+\lambda^{n+1}}{2(n+1)} \sum_{x\in\mathcal{X}}\sum_{y\in\mathcal{Y}} \rho(x,y)\left|\rho(x,y)-p(x)q(y)\right|^{n+1}$$

$$\times\left|\frac{1}{p^{n+1}(x)q^{n+1}(x)} - \frac{1}{\rho^{n+1}(x,y)}\right|.$$

Proof. Choose $a = p(x)q(y)$ and $h = \rho(x,y)$ and $b = \rho(x,y)$ to give, from (4.5), after multiplication by $\rho(x,y)$ and summation over $x \in \mathcal{X}$ and $y \in \mathcal{Y}$, the result as stated (4.11) follows.

5. Concluding Remarks

A Čebyšev-type functional introduced in [3] has been applied to obtain perturbed generalised Taylor series together with sharp bounds of the approximations. The approximation of the logarithmic function is given as an example which is further used to estimate the Kullback-Leibler distance, Shannon entropy and mutual information. These are of importance in information theory and, in particular, coding theory.

Acknowledgement. The author would like to thank Professor S. S. Dragomir for bringing the paper [9] to his attention and for the useful discussions.

References

1. P. Appell, *Sur une classe de polnômes*, Ann. Sci. Ecole Norm. Sup. **9(2)** (1880 119–144).
2. P. Cerone, *Generalised Taylor's formula with estimates of the remainder*, in Inequalities Theory and Applications, Vol. 2 (Y. J. Cho, J. K. Kim and S. S. Dragomir, Eds.), Nova Sciences Publ. Inc., 2003 [ONLINE: `http://rgmia.vu.edu.au/v5n2.html`], pp. 33–52.
3. ———, *Perturbed generalised Taylor's formulae with sharp bounds*, RGMIA Res. Rep. Coll. **5(4)** (2002), Article [ONLINE `http://rgmia.vu.edu.au/v5n4.html`].
4. P. Cerone and S. S. Dragomir, *New bounds for a perturbed generalised Taylor's formula*, East Asian Math. J. **17(2)** (2001), 197–215.
5. T. M. Cover and J. A. Thomas, *Elements of Information Theory*, John Wiley & Son Inc., 1991.
6. S. S. Dragomir, *Some integral inequalities of Grüss type*, Indian J. of Pure and Appl. Math. **31(4)** (2000), 397–415.
7. ———, *New estimation of the remainder in Taylor's formula using Grüss type inequalities and applications*, Math. Ineq. & Appl. **2(2)** (1999), 183–193.
8. ———, *An improvement of the remainder estimate in the generalised Taylor's formula*, RGMIA Res. Rep. Coll. **3(1)** (2000), Article 1.
9. S. S. Dragomir and V. Gluščević, *New estimates of the Kullback–Leibler distance and applications*, Inequality Theory and Applications–Volume 1 (Y. J. Cho, J. K. Kim and S. S. Dragomir, Eds.), Nova Science Publishers, Inc., Huntington, New York, 2001.
10. A. M. Fink, *A treatise on Grüss inequality*, Analytic and Geometric Inequalities and Applications, Math. Appl., Vol 478, Kluwer Academic Publishers, Dordrecht, 1999, pp. 93–114.
11. M. Matić, J. E. Pečarić and N. Uhević, *On new estimation of the remainder in generalised Taylor's formula*, Math. Ineq. & Appl. **2(3)** (1999), 343–361.
12. D. S. Mitrinović, J. E. Pečarić and A. M. Fink, *Classical and New Inequalities in Analysis*, Kluwer Academic Publishers, 1993.
13. J. E. Pečarić, F. Prochan and Y. L. Tong, *Convex Functions, Partial Orderings and Statistical Applications*, Academic Press, 1992.

In: Inequality Theory and Applications, Volume 4
Editors: Yeol Je Cho et al., pp. 47–60

ISBN 1-59454-874-9

GENERALIZED GAMES AND GENERALIZED VECTOR QUASI-EQUILIBRIA IN G-CONVEX SPACES

XIE PING DING[1] AND FU QUAN XIA[1]

ABSTRACT. In this paper, some new existence theorems of maximal elements for correspondences and Equilibrium theorems for generalized games are first established in noncompact G-convex space. As applications, some new existence theorems of solutions for generalized vector quasi-equilibrium problems are obtained in noncompact G-convex spaces. A new and unifying approach to solving the generalized vector quasi-equilibria problems in G-convex spaces is given. These results improve, unify and generalize many well-known results in recent literatures.

1. Introduction

In mathematical economics, showing the existence of equilibrium is the main problem of investigating various classes of economic models and others. Recently, Chowdhury-Tarafdar-Tan [7] established some new equilibrium existence theorems for generalized games in G-convex spaces which generalize the corresponding results of Ding and Tarafdar [17] and Ding [8] in topological vector spaces and H-spaces to G-convex spaces.

Inspired and motivated by extensive applications of the vector variational inequality problems and vector equilibrium problems with set-valued mappings, various generalized vector variational inequality problem, generalized vector equilibrium problem and generalized vector quasi-equilibrium problem have become important developed directions of the classical vector variational inequality problem, see Giannessi [18,19] and the references therein.

Let X, Y be a topological space, $F : X \times X \to 2^Y$ $A : X \to 2^X$ and $C : X \to 2^Y$ be set-valued mappings where 2^Y denotes the family of all subsets of Y. We shall introduce and study a new class of generalized vector quasi-equilibrium problem (in short, GVQEP) which is to find $\hat{x} \in X$ such that

$$\hat{x} \in A(\hat{x}) \quad \text{and} \quad F(\hat{x}, y) \not\subseteq C(\hat{x}), \quad y \in A(\hat{x}). \tag{1.1}$$

If $Y = \mathbf{R}$, $C(x) = (0, +\infty)$ for all $x \in X$ and $F(x, y)$ is a single-valued mapping, then the GVQEP (1.1) reduces to the following quasi-equilibrium problem (in short, QEP): find $\hat{x} \in X$ such that

$$\hat{x} \in A(\hat{x}) \quad \text{and} \quad F(\hat{x}, y) \leq 0, \quad y \in A(\hat{x}). \tag{1.2}$$

Received October 31, 2003.
1991 *Mathematics Subject Classification.* 91A80, 91B50, 49J53.
Key words and phrases. Generalized game, generalized vector quasi-equilibrium problems, transfer compactly open-valued correspondence, G-convex space.
This projected was supported by the Natural Science Foundation of Sichuan Education Department of China.

[1] College of Mathematics and Software Science, Sichuan Normal University, Chengdu, Sichuan 610066, P. R. China. E-mail: dingxip@sicnu.edu.cn

The QEP (1.2) and its variations have been introduced and investigated by many authors, For example, see Lin-Park [21], Ding [9–11], Lin-Yu [24] and Chen-Lin-Park [8].

If $A(x) = X$ for all $x \in X$, then the GVQEP (1.1) reduces the following genaralized vector equilibrium problem (in short, GVEP) which is to find $\hat{x} \in X$ such that

$$(1.3) \qquad F(\hat{x}, y) \not\subseteq C(x), \quad y \in X.$$

The GVEP (1.3) and its various special cases have been introduced and studied by Ding-Park [13,14], Oettli-Schläger [25], Ansari-Yao [2] and Lin-Yu Kassay [23]. Some existence theorems of solutions for the GVEP (1.3) are proved under different assumptions of topological spaces, G-convex spaces and topological vector spaces.

Let X be a nonempty convex subset of a topological vector space and Y be a topological vector space. Let $P : X \to 2^Y$ is a set-valued mapping such that for each $x \in X$, $P(x)$ is a closed convex cone in Y with $\operatorname{int} P(x) \neq \emptyset$ and $P(x) \neq Y$. If $C(x) = -\operatorname{int} P(x)$ for each $x \in X$, then the GVEP (1.3) reduces to the following GVEP which is to find $\hat{x} \in X$ such that

$$(1.4) \qquad F(\hat{x}, y) \not\subseteq -\operatorname{int} P(\hat{x}), \quad y \in X.$$

The GVEP (1.4) and its special cases have been studied by Ansari-Oettli-Schläger [1], Oettli-Schläger [25-28], Song [32] and Ansari-Yao [2] and others, see Giannesi [19].

In this paper, we first established some fixed point theorem, existence theorem od maximal elements and equilibrium existence theorems of generalized games. As applications, some existence theorems of solutions for the GVQEP (1.1) are proved in noncompact G-convex spaces. These theorems improve, unify and generalize many known results in recent literature.

2. Preliminaries

Let X be a topological space. We shall denote by 2^X and $\mathcal{F}(X)$ the family of all subsets of X and the family of all nonempty finite subsets of X, respectively. $|X|$ denotes the cardinality of X. Let $\triangle_n$ be the standard n-simplex with vertices $e_0, e_1, \cdots, e_n$ where e_i is the $(i+1)th$ unit vector in R^{n+1}. For $A \subset X$, $\mathrm{cl}_X(A)$ and $\int_X(A)$ denote the closure and interior of A in X, respectively. A subset A of X is said to be compactly open (resp., compactly closed) if for each nonempty compact subset K of X, $A \cap K$ is open (resp., closed) in K. Define the compact interior and the compact closure of A, denoted by $\operatorname{cint} A$ and $\mathrm{ccl} A$ as

$$\operatorname{cint} A = \bigcup \{B \subset X : B \subseteq A, B \text{ is compactly open in } X\}.$$
$$\mathrm{ccl}\, A = \bigcap \{B \subset X : A \subset B, B \text{ is compactly closed in } X\}.$$

It is easy to see that for any nonempty compact subset K of X, we have

$$\operatorname{cint}(A) \cap K = \operatorname{int}_K(A \cap K), \quad \mathrm{ccl}(A) \cap K = \mathrm{cl}_K(A \cap K)$$

and hence $\operatorname{cint}(A)$ (respectively, $\mathrm{ccl}(A)$) is compactly open (respectively, compactly closed) in X. By the definitions, a subset A of X is compactly open (respectively, compactly closed) in X if and only if $A = \operatorname{cint} A$ (respectively, $\mathrm{ccl}A = A$). If $F : X \to 2^Y$ is a correspondence, for any $y \in Y$ and $B \subset Y$, let $F(X) = \bigcup_{x \in X} F(x)$, $F^{-1}(y) = \{x \in X : y \in F(x)\}$ and $F^{-1}(B) = \{x \in X : F(x) \cap B \neq \emptyset\}$. Let $S, T : X \to 2^Y$ be two correspondences, then $T \cap S : X \to 2^Y$ is a correspondence defined by $(T \cap S)(x) = T(x) \cap S(x)$ for each $x \in X$. A mapping $T : X \to 2^Y$ is said to be upper semicontinuous if $T^{-1}(M)$ is closed in X for each closed subset M of Y. $S : X \to 2^Y$ is said to have open lower sections if for each $y \in Y$, $S^{-1}(y)$ is open in X. S is said to have compactly open lower section if for each $y \in Y$ and for any nonempty compact subset K of X, $S^{-1}(y) \cap K$ is open in K. S is said to be transfer compactly open-valued (respectively, transfer compactly closed-valued) on X if for each $x \in X$ and each nonempty compact subset K of Y, $y \in S(x) \cap K$ (respectively, $y \notin S(x) \cap K$) implies that there exists $x' \in X$ such that $y \in \operatorname{cint} S(x') \cap K$ (respectively, $y \notin \operatorname{ccl} S(x') \cap K$).

The notion of a generalized convex (G-convex) space was introduced under a extra isotonic condition by Park-Kim [30], [31]. Recently, Park [29], by removed the extra condition, give the following definition of a G-convex space.

A G-convex space (X, D, Γ) consists of a topological space X, a nonempty set D and a set-valued mapping $\Gamma : \mathcal{F}(D) \to 2^X \setminus \{\emptyset\}$ such that for each $A = \{a_0, a_1, \cdots, a_n\} \in \mathcal{F}(D)$, there exists a continuous mapping $\phi_A : \triangle_n \to \Gamma(A)$ such that $J \subset \{0, 1, \cdots, n\}$ implies $\phi_A(\triangle_J) \subset \Gamma(\{a_j : j \in J\})$ where $\triangle_J = co\{e_j : j \in J\}$. When $D = X$, write $(X, X, \Gamma) = (X, \Gamma)$. In case $D \subset X$, a subset K of (X, D, Γ) is said to be G-convex if for each $A \in \mathcal{F}(D)$, $A \subset K$ implies $\Gamma(A) \subset K$. We defined the G-convex hull of K, denoted by $G\text{-}\operatorname{co}(K)$, as

$$G\text{-}\operatorname{co}(K) = \bigcap\{B \subset X : K \subset B \text{ and } B \text{ is } G\text{-convex}\}.$$

By Lemma 2.1 of Tan and Zhang [33], we have

$$G\text{-}\operatorname{co}(K) = \bigcup\{G\text{-}\operatorname{co}(A) : A \in \mathcal{F}(K)\}.$$

Major examples of G-convex space are convex subsets of topological vector spaces, Lassonde's convex spaces, C-spaces (H-spaces) due to Horvath, Metric spaces with Micheal's convex structure, Pasicki's S-contractible space, Horvath's pseudoconvex spaces, Komiya's convex space, Bielawski's simplicial convexity spaces, Joo's pseudoconvex spaces, Hyperconvex metric spaces and so on, see [29]-[31].

Let $(X; \Gamma)$ be a G-convex space and Y be a real topological vector space. Let $F, G : X \times X \to 2^Y$ and $P : X \to 2^Y$ be set-valued mappings such that for each $x \in X$, $P(x)$ is a closed convex cone with $P(x) \neq Y$ and $\int P(x) \neq \emptyset$. Following Oettli-Schläger [28], $F(x, y)$ is said to be G-monotone if for all $x, y \in X$,

$$F(x, y) \nsubseteq -\operatorname{int} P(x) \;\Rightarrow\; G(x, y) \subseteq -P(x).$$

$F(x, y)$ is said to be maximal G-monotone if F is G-monotone and for all $x \in X$,

$$G(x, y) \subseteq -P(x),\; y \in X \;\Rightarrow\; F(x, y) \nsubseteq -\operatorname{int} P(x),\; y \in X.$$

Now we introduce a generalized convexity notion for set-valued mappings defined on a G-convex space. A set-valued mapping $F : X \times X \to 2^Y$ is said to be G-P_x-quasiconvex in y if for each finite set $N = \{y_1, \cdots, y_n\}$ in X and for each $y \in \Gamma(N)$, there exists $\alpha_1, \cdots, \alpha_n \in \mathbf{R}$ with $\alpha_i \geq 0$ and $\sum_{i=1}^n \alpha_i = 1$ such that $F(x, y) \subseteq \sum_{i=0}^n \alpha_i F(x, y_i) - P(x)$. If X is a nonempty convex subset of a topological vector space, then $F(x, y)$ is said to be P_x-quasiconvex in y if for each finite set $N = \{y_1, \cdots, y_n\}$ in X and for each $y \in \operatorname{co}(N)$ (i.e., $y = \sum_{i=1}^n \alpha_i y_i$ with $\alpha_i \geq 0$ and $\sum_{i=1}^n \alpha_i = 1$), $F(x, y) \subseteq \sum_{i=1}^n \alpha_i F(x, y_i) - P(x)$. Obviously, the above notion generalizes the corresponding notions of P_x-quasiconvexity-like introduced by Ansari and Yao [2].

Lemma 2.1. *Let X be a topological space and (Y, Γ) be a G-convex space. Suppose the mapping $\varphi : X \to 2^Y \setminus \{\emptyset\}$ is such that φ^{-1} is transfer compactly open-valued on Y.*

Then the mapping $\psi : X \to 2^Y \setminus \{\emptyset\}$ *defined by* $\psi(x) = G\text{-co}(\varphi(x))$ *is such that* ψ^{-1} *is also transfer compactly open-valued on* Y.

Proof. Let $y \in Y$ and K be any nonempty compact subset of X. By Lemma 2.1 of Tan and Zhang [33], if $x \in \psi^{-1}(y) \cap K$, then $x \in K$ and $y \in \psi(x) = G\text{-co}(\varphi(x)) = \cup\{G\text{-co}(A) : A \in \mathcal{F}(\varphi(x))\}$. There exists $A = \{a_1, \cdots, a_n\} \in \mathcal{F}(\varphi(x))$ such that $y \in G\text{-co}(A)$. It follows that $x \in \bigcap_{i=1}^n (\varphi^{-1}(a_i) \cap K)$ and so for each $i \in \{1, \cdots, n\}$, $x \in \varphi^{-1}(a_i) \cap K$. Since φ^{-1} is transfer compactly open-valued on Y, there exists $y_i \in Y$ such that $x \in \text{cint}(\varphi^{-1}(y_i)) \cap K$. It follows that $x \in \bigcap_{i=1}^n (\text{cint}\,\varphi^{-1}(y_i)) \cap K$. Clearly, $\bigcap_{i=1}^n (\text{cint}\,\varphi^{-1}(y_i)) \cap K$ is an open neighbourhood of x in K. Let $U = \cap_{i=1}^n (\text{cint}\,\varphi^{-1}(y_i)) \cap K$. Then for each $z \in U$, we have

$$z \in \text{cint}(\varphi^{-1}(y_i)) \subseteq \varphi^{-1}(y_i), \quad i = 1, \cdots, n.$$

Now if we choice a $\hat{y} \in G\text{-co}(\{y_1, \cdots, y_n\})$, then we have $\hat{y} \in G\text{-co}(\{y_1, \cdots, y_n) \subseteq G\text{-co}(\varphi(z)) = \psi(z)$ and $z \in \psi^{-1}(\hat{y}) \cap K$. Therefore $U \subseteq \psi^{-1}(\hat{y}) \cap K$ and $x \in U \subseteq \psi^{-1}(\hat{y}) \cap K$. It follows that $\psi^{-1}(\hat{y}) \cap K$ is an open set in K. Then we have $x \in \psi^{-1}(\hat{y}) \cap K = \text{cint}(\psi^{-1}(\hat{y}) \cap K) \subseteq \text{cint}(\psi^{-1}(\hat{y})) \cap K$. Therefore ψ^{-1} is transfer compactly open-valued on Y. This completes the proof.

Remark 2.1. Lemma 2.1 generalizes Lemma 3.1 of Ding [8], Lemma 2.2 of Tan-Zhang [33] and Lemma 2.1 of Chowdhury-Tarafdar-Tan [7] to the correspondence without compactly open lower sections

The following result is a special case of Theorem 3.4 of Ding [12].

Lemma 2.2. *Let* X *be a nonempty subset of a* G*-convex space* $(E, D; \Gamma)$, K *be a nonempty compact subset of* X *and* $G : X \to 2^E$ *be such that*

(i) G *is compactly closed-valued,*

(ii) *for each* $A \in \mathcal{F}(X \cap D), \Gamma(A) \subseteq \bigcup_{x \in A} G(x)$,

(iii) *for each* $N \in \mathcal{F}(X \cap D)$, *there exists a compact* G*-convex subset* L_N *of* E *containing* N *such that for each* $y \in L_N \setminus K$, *there is a* $x \in L_N \cap X$ *satisfying* $y \notin G(x)$.

Then $K \cap (\bigcap_{x \in X} G(x)) \neq \emptyset$.

Lemma 2.3. [3] *Let* X *and* Y *be topological spaces. If* $T : X \to 2^Y$ *is an upper semicontinuous mapping with closed values, the* T *has closed graph.*

The following result is Lemma 3.1 of Chowdhury-Tarafdar-Tan [7] which is a variant form of Lemma 5.1 of Ding [8] in G-convex spaces

Lemma 2.4. *Let* (X, Γ) *be a regular* G*-convex space and* Y *be a non-empty subset of a* G*-convex space* (E, Γ). *Let* $\theta : X \to E$ *and* $P : X \to 2^Y$ *such that:*

(1) *each open subset of* X *which contains the set* $B = \{x \in X : P(x) \neq \emptyset\}$ *is paracompact,*

(2) *for each* $x \in X$ *with* $P(x) \neq \emptyset$, *there exist* $\varphi_x, \psi_x : X \to 2^Y$ *and an open neighbourhood* N_x *of* x *in* X *such that*

(a) *for each* $z \in N_x$, $P(z) \subseteq \varphi_x(z)$ *and* $\theta(z) \notin G\text{-co}(\varphi_x(z))$,

(b) *for each* $z \in X$, $\psi_x(z) \subseteq \varphi_x(z)$ *and* $G\text{-co}(\varphi_x(z) \subseteq Y$,

(c) *for each* $y \in Y$, $\psi_x^{-1}(y)$ *is compactly open in* X,

(d) *for any non-empty finite subset* A *of the set* $\{x \in X : P(x) \neq \emptyset\}$, *we have*

$$\Big\{z \in \bigcap_{x \in A} N_x : \bigcap_{x \in A} G\text{-co}(\varphi_x(z)) \neq \emptyset\Big\}$$
$$= \Big\{z \in \bigcap_{x \in A} N_x : \bigcap_{x \in A} G\text{-co}(\psi_x(z)) \neq \emptyset\Big\}.$$

Then there exist correspondences $\varphi, \psi : X \to 2^Y$ *such that*

(I) $P(x) \subseteq \varphi(x)$ *for each* $x \in X$,

(II) *for each* $x \in X$, $G\text{-co}(\varphi(x)) \subseteq Y$ *and* $\theta(x) \notin G\text{-co}(\varphi(x))$,

(III) *for each* $x \in X$, $\psi(x) \subseteq \varphi(x)$ *and for each* $y \in Y$, $\psi^{-1}(y)$ *is compactly open in* X *and* $\{x \in X : \varphi(x) \neq \emptyset\} = \{x \in X : \psi(x) \neq \emptyset\}$.

The following result is Lemma 2.2 of Ding-Park [23].

Lemma 2.5. *Let* X *and* Y *be two topological spaces,* $F : X \times X \to 2^Y$ *and* $C : X \to 2^Y$ *be two set-valued mappings such that*

(i) *the mapping* $C(\cdot)$ *has open graph in* $X \times Y$,

(ii) *for each* $y \in X$, *the mapping* $x \mapsto g(x, y)$ *is upper semicontinuous on each compact subset of* X *with nonempty compact values.*

Then for each $y \in X$, *the set* $\{x \in X : F(x, y) \subseteq C(x)\}$ *is compactly open in* X.

3. Maximal Elements and Equilibrium Existence

Theorem 3.1. *Let* X *be a nonempty* G*-convex subset of a* G*-convex space* (E, Γ), K *be a nonempty compact subset of* X. *Let* $Q,\ P : X \to 2^X$ *be such that*

(1) *for each* $x \in X$, $P(x) \subseteq Q(x)$,

(2) P^{-1} *is transfer compactly open-valued in* X,

(3) *for each* $N \in \mathcal{F}(X)$, *there exists a compact* G*-convex subset* L_N *of* X *containing* N *such that for each* $y \in L_N \setminus K$, *there is an* $x \in L_N$ *satisfying* $y \in \text{cint}(G\text{-co}\, Q)^{-1}(x)$,

(4) *for each* $x \in K$, $P(x) \neq \emptyset$.

Then there exists a point $\hat{x} \in X$ *such that* $\hat{x} \in G\text{-co}(Q(\hat{x}))$.

Proof. Suppose that the conclusion does not hold. Then for each $x \in X$, $x \notin G\text{-co}(Q(x))$. By condition (2) and Lemma 2.1, $(G\text{-co}\, P)^{-1}$ is also transfer compactly open-valued on X. Define a correspondence $T, S : X \to 2^X$ by

$$T(x) = \text{ccl}[X \setminus (G\text{-}\operatorname{co} Q)^{-1}(x)], \quad x \in X,$$
$$S(x) = \text{ccl}[X \setminus (G\text{-}\operatorname{co} P)^{-1}(x)], \quad x \in X.$$

By condition (1), it is easy to know that for each $x \in X$, $T(x) \subset S(x)$. We will prove that T satisfies all conditions of Lemma 2.2. Clearly, T is compactly closed values. We claim that for any $A \in \mathcal{F}(X)$, $\Gamma(A) \subseteq \bigcup_{x \in A} T(x)$. If it is not true, then there exist $A = \{x_1, \cdots, x_n\} \in \mathcal{F}(X)$, and $y \in \Gamma(A)$ such that $y \notin \bigcup_{x \in A} T(x)$. Therefore $y \notin T(x)$ for all $x \in A$ and hence $y \notin X \setminus (G\text{-co}\, Q)^{-1}(x)$ for all $x \in A$. It follows that

$$A \subseteq G\text{-}\operatorname{co} Q(y), \quad y \in \Gamma(A) \subseteq G\text{-}\operatorname{co} Q(y),$$

which is a contradiction. Hence we have that for any $A \in \mathcal{F}(X)$, $\Gamma(A) \subseteq \bigcup_{x \in A} T(x)$. By condition (3), for each $N \in \mathcal{F}(X)$, there exists a compact G-convex subset L_N of X containing N such that for each $y \in L_N \setminus K$, there is an $x \in L_N$ satisfying $y \in \text{cint}(G\text{-co}\, Q)^{-1}(x)$. It follows that

$$y \notin X \setminus \text{cint}((G\text{-}\operatorname{co} Q)^{-1}(x)) = \text{ccl}[X \setminus (G\text{-}\operatorname{co} Q)^{-1}(x)] = T(x).$$

By Lemma 2.2 with $X = E$, $K \cap \bigcap_{x \in X} T(x) \neq \emptyset$. Since $T(x) \subset S(x)$ for each $x \in X$, then we have $K \cap \bigcap_{x \in X} S(x) \neq \emptyset$. That is,

$$K \cap \bigcap_{x \in X} \text{ccl}[X \setminus (G\text{-}\operatorname{co} P)^{-1}(x)] \neq \emptyset. \tag{3.1}$$

Define a correspondence $G : X \to 2^X$ by

$$G(x) = X \setminus (G\text{-}\operatorname{co} P)^{-1}(x), \quad x \in X.$$

Then G is transfer compactly closed-valued on X. By (3.1), we have

$$K \cap \bigcap_{x \in X} \operatorname{ccl} G(x) \neq \emptyset.$$

It follows that

$$\bigcap_{x \in X} \operatorname{cl}_K(G(x) \cap K) = \bigcap_{x \in X} (\operatorname{ccl}(G(x) \cap K) = K \cap \bigcap_{x \in X} \operatorname{ccl} G(x) \neq \emptyset.$$

Now we shall prove that

$$\bigcap_{x \in X} \operatorname{cl}_K(G(x) \cap K) = \bigcap_{x \in X} (G(x) \cap K) = K \cap \bigcap_{x \in X} G(x).$$

Clearly, $\bigcap_{x \in X}(G(x) \cap K) \subseteq \bigcap_{x \in X} \operatorname{cl}_K(G(x) \cap K)$. So we only need to show $\bigcap_{x \in X} \operatorname{cl}_K(G(x) \cap K) \subseteq \bigcap_{x \in X}(G(x) \cap K)$. Suppose it is false, then there exists an $y \in \bigcap_{x \in X} \operatorname{cl}_K(G(x) \cap K)$ such that $y \notin \bigcap_{x \in X}(G(x) \cap K)$ and so $y \notin G(x) \bigcap K$ for some $x \in X$. Since G is transfer compactly closed values, there exists an $x' \in X$ such that $y \notin \operatorname{ccl} G(x') \cap K = \operatorname{cl}_K(G(x') \cap K)$, which is a contradiction. Hence we must have $K \cap \bigcap_{x \in X} G(x) \neq \emptyset$. Thus we get

$$\begin{aligned} \emptyset \neq K \cap \Big(\bigcap_{x \in X} (X \setminus (G\text{-}\operatorname{co} P)^{-1}(x)) \Big) &= K \cap \Big(X \setminus \bigcup_{x \in X} (G\text{-}\operatorname{co} P)^{-1}(x) \Big) \\ &= K \setminus \bigcup_{x \in X} (G\text{-}\operatorname{co} P)^{-1}(x) \cap K). \end{aligned}$$

On the other hand, by condition (4), we know $P(x) \neq \emptyset$ for each $x \in K$, then G-co $P(x) \neq \emptyset$ for each $x \in K$. It follows that

$$K = \bigcup_{x \in X} ((G\text{-}\operatorname{co} P)^{-1}(x) \cap K),$$

which is a contradiction. Hence there must exist a point $\hat{x} \in K$ such that $\hat{x} \in G$-co $Q(\hat{x})$. This completes the proof.

Remark 3.1. (a) Theorem 3.1 generalizes Theorem 2.4 of Chowdhury-Tarafdar-Tan [7] in following ways:

(1) $P^{-1} : X \to 2^X$ may not be compactly open-valued,

(2) the assumptions that for each $x \in E$, $\Gamma(\{x\}) = \{x\}$ and for each compact G-convex subset A of X and for each $B \in \mathcal{F}(E)$, G-co$(A \cup B)$ is compact, and $\Gamma(B) = G$-co(B) are all dropped.

(b) If for each $y \in X$, $F^{-1}(y)$ is open or compactly open, or F^{-1} is transfer open-valued in X, then F^{-1} must be transfer compactly open-valued in X. Hence Theorem 3.2 generalize Theorem 3.1 of Ding and Tarafdar [17], Theorem 3.1 of Ding [8], Theorem 1 of Ding and Tan [15], Theorem 1 of Lin-Yu [22], Theorem 2.2 of Yuan [39] and Theorem 3 of Ding-Tan [16].

Theorem 3.2. *Let X be a non-empty G-convex subset of a G-convex space (E, Γ), K be a nonempty compact subset of X and $Q : X \to 2^X$ be such that*

(1) *for each $x \in X$, $x \notin G\text{-co}(Q(x))$,*

(2) *there exists a correspondence $\psi : X \to 2^Y$ such that for each $x \in X$, $\psi(x) \subseteq Q(x)$ and for each $y \in Y$, ψ^{-1} is transfer compactly open-valued in X and $\{x \in X : Q(x) \neq \emptyset\} = \{x \in X : \psi(x) \neq \emptyset\}$,*

(3) *for each $N \in \mathcal{F}(X)$, there exists a compact G-convex subset L_N of X containing N such that for each $y \in L_N \setminus K$, there is an $x \in L_N$ satisfying $y \in \text{cint}((G\text{-co}\, Q)^{-1}(x))$.*
Then there exists a point $\hat{x} \in K$ such that $Q(\hat{x}) = \emptyset$, i.e., $\hat{x}$ is a maximal element of Q.

Proof. Suppose $Q(x) \neq \emptyset$ for each $x \in K$. By condition (2), we have $\psi(x) \neq \emptyset$ for each $x \in K$. It follows from Theorem 3.1 that there exists $\hat{x} \in X$ such that $\hat{x} \in G\text{-co}(Q(\hat{x}))$, which contradicts condition (1). Hence there exists an $\hat{x} \in K$ such that $Q(\hat{x}) = \emptyset$.

Remark 3.2. Theorem 3.2 generalizes Theorem 4.1 of Ding-Tarafdar [17], Theorem 5.1 of Ding [8], Theorem 3.1 of Chowdhury-Tarafdar-Tan [7], Theorem 3.2 of Tan-Yuan [34], and Theorems 3 and 4 of Ding-Tan [15].

Theorem 3.3. *Let X be a non-empty paracompact G-convex subset of a G-convex space (E, Γ), K be a nonempty compact subset of X and $P : X \to 2^X$ be a set-valued mapping. Suppose the the following condition are satisfied:*

(1) *for each $x \in X$ with $P(x) \neq \emptyset$, there exist φ_x, $\psi_x : X \to 2^X$ and an open neighbourhood N_x of x in X such that*

(a) *for each $z \in N_x$, $P(z) \subset \varphi_x(z)$ and $z \notin G\text{-co}(\varphi_x(z))$,*

(b) *for each $z \in X$, $\psi_x(z) \subset \varphi_x(z)$,*

(c) *for each $y \in X$, $\psi_x^{-1}(y)$ is compactly open in X,*

(d) *for any non-empty finite subset A of the set $\{x \in X : P(x) \neq \emptyset\}$,*

$$\Big\{z \in \bigcap_{x\in A} N_x : \bigcap_{x\in A} G\text{-co}(\varphi_x(z)) \neq \emptyset\Big\}$$
$$= \Big\{z \in \bigcap_{x\in A} N_x : \bigcap_{x\in A} G\text{-co}(\psi_x(z)) \neq \emptyset\Big\}.$$

(2) *for each $N \in \mathcal{F}(X)$, there exists a compact G-convex subset L_N of X containing N such that for each $y \in L_N \setminus K$, there is an $x \in L_N$ satisfying $y \in \text{cint}((G\text{-co}\, P)^{-1}(x))$.*
Then there exists a point $\hat{x} \in K$ such that $P(\hat{x}) = \emptyset$, i.e., $\hat{x}$ is a maximal element of P.

Proof. Suppose that the conclusion does not hold. Then $P(x) \neq \emptyset$ for all $x \in X$ and hence the set $\{x \in X : P(x) \neq \emptyset\} = X$ is paracompact. By Lemma 2.4, there exist correspondences $Q, \psi : X \to 2^X$ such that

(I) $P(x) \subset Q(x)$ for each $x \in X$,

(II) for each $x \in X$, $x \notin G\text{-co}(Q(x))$,

(III) for each $x \in X$, $\psi(x) \subset Q(x)$ and for each $y \in Y$, $\psi^{-1}(y)$ is compactly open in X, and $\{x \in X : Q(x) \neq \emptyset\} = \{x \in X : \psi(x) \neq \emptyset\}$.

Note that for each $y \in L_N \setminus K$, there is an $x \in L_N$ satisfying

$$y \in \text{cint}((G\text{-}\operatorname{co} P)^{-1}(x)) \subset \text{cint}((G\text{-}\operatorname{co} Q)^{-1}(x)).$$

By Theorem 3.2, there exists a point $\hat{x} \in K$ such that $Q(\hat{x}) = \emptyset$, so that $P(\hat{x}) = \emptyset$, which is contradiction. Therefore, there exists a point $\hat{x} \in K$ such that $P(\hat{x}) = \emptyset$. This completes the proof.

Remark 3.3. Theorem 3.3 generalizes Theorem 3.2 of Chowdhury-Tarafdar-Tan. [7] in following way: the assumptions that for each $x \in E$, $\Gamma(\{x\}) = \{x\}$ and for each compact, G-convex subset A of X and for each $B \in \mathcal{F}(E)$, $G\text{-co}(A \cup B)$ is compact and $\Gamma(B) = G\text{-co}(B)$ are all dropped. Theorem 3.3 also unify and generalizes Theorem 4.2 of Ding-Tarafdar [17], Theorem 5.3 of Ding [8], Theorem 3.2 of Yuan [39], Theorem 3.3 of Tan-Yuan [34], Theorem 1 of Ding-Tan [16], Theorem 5 of Ding-Tan [15], Corollary 1 of Borglin-Keiding [5], Theorem 2 of Toussaint [35], Theorem 2 of Tulcea [36], Theorem 5.1 and Corollary 5.1 of Yannelis-Prabhakar [38] and Theorem 2 of Yannelis [37].

Theorem 3.4. *Let X be a non-empty G-convex subset of a G-convex space (E, Γ). Let A, B, $P : X \to 2^X$ such that:*

(1) for each $x \in X$, $A(x) \neq \emptyset$ and $G\text{-co}(A(x)) \subseteq \overline{B}(x)$,

(2) for each $y \in X$, $A^{-1}(y)$ is compactly open in X,

(3) $A \cap P : X \to 2^X$ is such that

(a) for each $x \in X$, $x \notin G\text{-co}((A \cap P)(x))$,

(b) there exists a correspondence $\varphi : X \to 2^Y$ such that for each $x \in X$, $\varphi(x) \subseteq (A \cap P)(x)$ and for each $y \in Y$, $\varphi^{-1}(y)$ is compactly open in X, and $\{x \in X : (A \cap P)(x) \neq \emptyset\} = \{x \in X : \varphi(x) \neq \emptyset\}$,

(4) for each $N \in \mathcal{F}(X)$, there exists a compact G-convex subset L_N of X containing N such that for each $y \in L_N \setminus K$, there is an $x \in L_N$ satisfying $y \in (G\text{-co}(A \cap P))^{-1}(x)$. Then there exists a point $\hat{x} \in K$ such that $\hat{x} \in \overline{B}(\hat{x})$ and $A(\hat{x}) \cap P(\hat{x}) = \emptyset$.

Proof. Let $M = \{x \in X : x \notin \overline{B}(x)\}$, then M is open in X. Define φ, $\psi : X \to 2^X$ by

$$\varphi(x) = \begin{cases} A(x) \cap P(x), & \text{if } x \notin M, \\ A(x), & \text{if } x \in M. \end{cases}$$

$$\psi(x) = \begin{cases} \varphi(x), & \text{if } x \notin M, \\ A(x), & \text{if } x \in M. \end{cases}$$

Then, by condition (3), for each $x \in X$, $\psi(x) \subseteq \varphi(x)$ and $\{x \in X : \psi(x) \neq \emptyset\} = \{x \in X : \varphi(x) \neq \emptyset\}$. For each $y \in X$, it is easy to see that $\psi^{-1}(y) = (M \cup \varphi^{-1}(y)) \cap A^{-1}(y)$ and it is compactly open in X by the conditions (2) and (3). For each $x \in X$, if $x \in M$, then $x \notin \overline{B}(x)$ and it follows from (1) that $x \notin G\text{-co}(A(x)) = G\text{-co}(\varphi(x))$; if $x \notin M$, then $x \notin G\text{-co}(A(x) \cap P(x)) = G\text{-co}(\varphi(x))$ by the condition (3)(a). By condition (4) and the definition of φ, for each $N \in \mathcal{F}(X)$, there exists a compact G-convex subset L_N of X containing N such that for each $y \in L_N \setminus K$, there is an $x \in L_N$ satisfying $y \in (G\text{-co}(A \cap P))^{-1}(x) = \text{cint}((G\text{-co}(A \cap P))^{-1}(x)) \subseteq \text{cint}((G\text{-co}\,\phi)^{-1}(x))$ since $(G\text{-co}(A \cap P))^{-1}$ is compactly open-valued. By Theorem 3.2, there exists a point $\hat{x} \in K$ such that $\varphi(\hat{x}) = \emptyset$. Since for each $x \in K$, $A(x) \neq \emptyset$, then we must have $\hat{x} \in \overline{B}(\hat{x})$ and $A(\hat{x}) \cap P(\hat{x}) = \emptyset$. This completes the proof.

Remark 3.4. Theorem 3.4 generalizes Theorem 4.1 of Chowdhury-Tarafdar-Tan [7] in following way:

the assumptions that for each $x \in E$, $\Gamma(\{x\}) = \{x\}$ and for each compact, G-convex subset A of X and for each $B \in \mathcal{F}(E)$, $G\text{-co}(A \bigcup B)$ is compact, and $\Gamma(B) = G\text{-co}(B)$ are all dropped. Theorem 3.3 also unify and generalizes Theorem 5.1 of Ding and Tarafdar [17], Theorem 6.1 of Ding [8], Theorem 4.1 of Tan-Yuan [34], Theorem 6 of Ding-Tan [15] and Theorem 4.1 of Yuan [39].

As a direct consequence of Theorem 3.4, we easily obtain the following result.

Corollary 3.1. *Let X be a non-empty G-convex subset of a G-convex space (E, Γ), K be a nonempty compact subset of X. Let A, B, $P : X \to 2^X$ such that:*

(1) for each $x \in X$, $A(x) \neq \emptyset$ and $G\text{-co}(A(x)) \subset \overline{B}(x)$,

(2) *for each* $y \in X$, $A^{-1}(y)$ *is compactly open in* X,

(3) *for each* $y \in Y$, $P^{-1}(y)$ *is compactly open in* X,

(4) *for each* $x \in X$, $x \notin G\text{-co}(P(x))$,

(5) *for each* $N \in \mathcal{F}(X)$, *there exists a compact* G*-convex subset* L_N *of* X *containing* N *such that for each* $y \in L_N \setminus K$, *there is an* $x \in L_N$ *satisfying* $y \in (G\text{-co}(A \cap P))^{-1}(x)$.

Then there exists a point $\hat{x} \in K$ *such that* $\hat{x} \in \overline{B}(\hat{x})$ *and* $A(\hat{x}) \cap P(\hat{x}) = \emptyset$.

4. Existence of Solutions For the GVQEP

Theorem 4.1. *Let* Y *be a nonempty set,* $(X;\Gamma)$ *be a* G*-convex space and* K *be a nonempty compact subset of* X. *Let* $F, G : X \times X \to 2^Y, C, D : X \to 2^Y$ *and* $A : X \to 2^X$ *be set-valued mappings. Suppose that that following conditions are satisfied:*

(1) $A : X \to 2^X$ *has nonempty closed* G*-convex values,*

(2) $A^{-1} : X \to 2^X$ *is compactly open-valued,*

(3) *for all* $x \in X$, $G(x,x) \not\subseteq D(x)$,

(4) *for each* $y \in X$, *the set* $P(x) = \{x \in X : F(x,y) \subseteq C(x)\}$ *is compactly open in* X,

(5) *for all* $x, y \in X$, $F(x,y) \subseteq C(x)$ *implies* $G(x,y) \subseteq D(y)$,

(6) *for each* $x \in X$, *the set* $H(x) = \{y \in X : G(x,y) \subseteq D(y)\}$ *is* G*-convex,*

(7) *for each* $N \in \mathcal{F}(X)$, *there exists a compact* G*-convex subset* L_N *of* X *containing* N *such that for each* $y \in L_N \setminus K$, *there is an* $x \in L_N$ *satisfying* $y \in (G\text{-co}(A \cap P))^{-1}(x)$ *where* $P(x) = \{y \in Y : F(x,y) \subseteq C(x)\}$ *for each* $x \in X$.

Then there exists an $\hat{x} \in K$ *such that*

$$\hat{x} \in A(\hat{x}) \quad \text{and} \quad F(\hat{x}, y) \not\subseteq C(\hat{x}), \quad y \in A(\hat{x}).$$

Proof. By (1) and (2), A has nonempty closed G-convex values and for each $y \in X$, $A^{-1} : X \to 2^X$ is compactly open-valued, then the conditions (1) and (2) of Corollary 3.1 with $A = B$ are satisfied. By conditions (4) and (7), It is easy to see that the conditions (3) and (5) of Corollary 3.1 with $A = B$ are satisfied. By the condition (3) and (6), for each $x \in X$, $G(x,x) \not\subseteq D(x)$ and the set $H(x) = \{y \in X : G(x,y) \subseteq D(y)\}$ is G-convex. It follows that for each $x \in X$, $x \notin H(x) = G\text{-co}(H(x))$. By condition (5), we have

$$P(x) = \{y \in X : F(x,y) \subseteq C(x)\} \subseteq \{y \in X : G(x,y) \subseteq D(y)\} = H(x), \quad x \in X.$$

It follows that for each $x \in X$, $G\text{-co}(P(x)) \subseteq G\text{-co}(H(x)) = H(x)$ and so $x \notin G\text{-co}(P(x))$. All conditions of Corollary 3.1 with $A = B$ are satisfied, by Corollary 3.1, there exists an $\hat{x} \in K$ such that $\hat{x} \in \overline{A}(\hat{x})$ and $A(\hat{x}) \cap P(\hat{x}) = \emptyset$. Note that A has closed values, we obtain that $\hat{x} \in A(\hat{x})$ and $A(\hat{x}) \cap P(\hat{x}) = \emptyset$. Therefore there exists an $\hat{x} \in K$ such that

$$\hat{x} \in A(\hat{x}) \quad \text{and} \quad F(\hat{x}, y) \not\subseteq C(\hat{x}), \quad \forall y \in A(\hat{x}).$$

Theorem 4.2. *Let* Y *be a nonempty set,* $(X;\Gamma)$ *be a* G*-convex space and* K *be a nonempty compact subset of* X. *Let* $F, G : X \times X \to 2^Y, C, D : X \to 2^Y$ *and* $A : X \to 2^X$ *be set-valued mappings. Suppose that that following conditions are satisfied:*

(1) $A : X \to 2^X$ *has nonempty closed* G*-convex values,*

(2) $A^{-1} : X \to 2^X$ *is compactly open-valued,*

(3) *for all* $x \in X$, $G(x,x) \not\subseteq D(x)$,

(4) *for each* $y \in X$, *the set* $P(x) = \{x \in X : F(x,y) \subseteq C(x)\}$ *is compactly open in* X,

(5) *for all* $x, y \in X$, $F(x,y) \subseteq C(x)$ *implies* $G(x,y) \subseteq D(x)$,

(6) *for each* $x \in X$, *the set* $H(x) = \{y \in X : G(x,y) \subseteq D(x)\}$ *is* G*-convex,*

(7) for each $N \in \mathcal{F}(X)$, there exists a compact G-convex subset L_N of X containing N such that for each $y \in L_N \setminus K$, there is an $x \in L_N$ satisfying $y \in (G\text{-co}(A \cap P))^{-1}(x)$.
Then there exists an $\hat{x} \in K$ such that $\hat{x} \in A(\hat{x})$ and $F(\hat{x}, y) \not\subseteq C(\hat{x})$ for all $y \in A(\hat{x})$.

Proof. By using a similar argument as in the proof of Theorem 4.1, It is easy to show that the conclusion of Theorem 4.2 holds.

If $A(x) = X$ for each $x \in X$, then condition (1) and (2) of Theorem 4.2 are satisfied trivially. Hence we have the following result.

Corollary 4.1. *Let Y be a nonempty set, $(X; \Gamma)$ be a G-convex space and K be a non-empty compact subset of X. Let $F, G : X \times X \to 2^Y$ and $C, D : X \to 2^Y$ be set-valued mappings. Suppose the following conditions are satisfied:*
(1) for all $x \in X$, $G(x, x) \not\subseteq D(x)$,
(2) for each $y \in X$, the set $P(x) = \{x \in X : F(x, y) \subseteq C(x)\}$ is compactly open in X,
(3) for all $x, y \in X$, $F(x, y) \subseteq C(x)$ implies $G(x, y) \subseteq D(x)$,
(4) for each $x \in X$, the set $\{y \in X : G(x, y) \subseteq D(x)\}$ is G-convex;
(5) for each $N \in \mathcal{F}(X)$, there exists a compact G-convex subset L_N of X containing N such that for each $y \in L_N \setminus K$, there is an $x \in L_N$ satisfying $y \in (G\text{-co}\, P)^{-1}(x)$.
Then there exists an $\hat{x} \in K$ such that

$$F(\hat{x}, y) \not\subseteq C(\hat{x}), \quad y \in X.$$

Remark 4.1. Note that Y may be any nonempty set without linear and topological structure in Theorems 4.1, 4.2 and Corollary 4.1. Corollary 4.1 is a improving variant from of Theorem 4.4 of Ding-Park [13] and Theorem 3.2 of Lin-Yu-Kassay [23]. Hence Theorem 4.2 generalizes Theorem 4.4 of Ding-Park [13] and Theorem 3.2 of Lin-Yu-Kassay [23] from the GVEP to the GVQEP.

Theorem 4.3. *Let $(X; \Gamma)$ be a G-convex space, K be a non-empty compact subset of X and Y be a topological space. Let $F, G : X \times X \to 2^Y$, $A : X \to 2^X$ and $C, D : X \to 2^Y$ be set-valued mapping. Suppose the following conditions are satisfied:*
(1) $A : X \to 2^X$ has nonempty closed G-convex values,
(2) $A^{-1} : X \to 2^X$ is compactly open-valued,
(3) for all $x \in X$, $G(x, x) \not\subseteq D(x)$,
(4) for all $x, y \in X$, $F(x, y) \not\subseteq C(x)$ implies $G(x, y) \subseteq D(x)$,
(5) for each $y \in X$, $x \mapsto F(x, y)$ is upper semi-continuous with compact values and $C : X \to Y$ has open graph,
(6) for each $x \in X$, the set $\{y \in X : G(x, y) \subseteq D(x)\}$ is G-convex,
(7) for each $N \in \mathcal{F}(X)$, there exists a compact G-convex subset L_N of X containing N such that for each $y \in L_N \setminus K$, there is an $x \in L_N$ satisfying $y \in (G\text{-co}(S \cap H))^{-1}(x)$., where $H : X \to 2^X$ is defined by $H(x) = \{y \in X : F(x, y) \subseteq C(x)\}$.
Then there exists an $\hat{x} \in K$ such that $\hat{x} \in A(\hat{x})$ and $F(\hat{x}, y) \not\subseteq -\operatorname{int} P(\hat{x})$ for all $y \in A(\hat{x})$.

Proof. From condition (5) and Lemma 2.5 it follows that for each $y \in X$, the set $\{x \in X : F(x, y) \subseteq C(x)\}$ is compact open in X, the condition (4) of Theorem 4.2 is satisfied. The conclusion of Theorem 4.3 holds from Theorem 4.2.

Remark 4.2. Theorem 4.3 generalizes Theorem 3.3 of Lin-Yu-Kassay [23] from the GVEP to the GVQEP.

Corollary 4.2. *Let $(X;\Gamma)$ be a G-convex space, K be a non-empty compact subset of X and Y be a topological vector space. Let $F, G : X \times X \to 2^Y$, $A : X \to 2^X$ and $P : X \to 2^Y$ such that for each $x \in X$, $P(x)$ is a closed convex cone with $P(x) \neq Y$ and $\operatorname{int} P(x) \neq \emptyset$. Suppose the following conditions are satisfied:*

(1) $A : X \to 2^X$ has nonempty closed G-convex values,

(2) $A^{-1} : X \to 2^X$ is compactly open-valued,

(3) for all $x \in X$, $G(x,x) \not\subseteq -\operatorname{int} P(x)$,

(4) for all $x, y \in X$, $F(x,y) \subseteq -\operatorname{int} P(x)$ implies $G(x,y) \subseteq -P(x)$,

(5) for each $y \in Y$, $x \mapsto F(x,y)$ is upper semi-continuous with compact values and $W : X \to 2^Y$ defined by $W(x) = Y \setminus (-\operatorname{int} P(x))$ is upper semicontinuous for each $x \in X$,

(6) $G(x,y)$ is G-P_x-quasi-convex in y,

(8) for each $N \in \mathcal{F}(X)$, there exists a compact G-convex subset L_N of X containing N such that for each $y \in L_N \setminus K$, there is an $x \in L_N$ satisfying $y \in (G\text{-co}(A \cap H))^{-1}(x)$, where $H : X \to 2^X$ is defined by $H(x) = \{y \in X : F(x,y) \subseteq -\operatorname{int} P(x)\}$.

Then there exists an $\hat{x} \in K$ such that $\hat{x} \in A(\hat{x})$ and $F(\hat{x}, y) \not\subseteq -\operatorname{int} P(\hat{x})$ for all $y \in A(\hat{x})$.

Proof. For each $x \in X$, let $C(x) = -\operatorname{int} P(x)$ and $D(x) = -P(x)$ for each $x \in X$. By the condition (5) and Lemma 2.3, it follows that W has closed graph, and hence $C : X \to 2^Y$ has an open graph. Since $x \mapsto F(x,y)$ is upper semicontinuous on each compact subset of X with compact values, It follows from Lemma 2.5 that for each $y \in X$, the set $\{x \in X : F(x,y) \subseteq -\operatorname{int} P(x)\}$ is compactly open in X. Now we prove that for each $x \in X$, the set $Q(x) = \{y \in X : G(x,y) \subseteq -\operatorname{int} P(x)\}$ is G-convex. For any $N = \{y_1, \cdots, y_n\} \in \mathcal{F}(Q(x))$ and any $y \in \Gamma(N)$, by the condition (6), there exist $\alpha_i \geq 0$ for $i = 1, \cdots, n$ with $\sum_{i=1}^n \alpha_i = 1$ such that

$$G(x,y) \subseteq \sum_{i=1}^{n} \alpha_i G(x, y_i) - P(x) \subseteq -\operatorname{int} P(x) - P(x) = -\operatorname{int} P(x).$$

It follows that $\Gamma(N) \subseteq Q(x)$. All condition of Theorem 4.2 are satisfied. The conclusion of Corollary 4.2 holds from Theorem 4.2. This completes the proof.

Remark 4.3. Corollary 4.2 generalizes Corollaries 4.4 and 4.5 of Ding-Park [13] and Corollary 3.2 of Lin-Yu-Kassay [23] from the GVEP to the GVQEP. Corollary 4.2 also generalizes Theorem 2.1 of Ansari and Yao [14] in the following ways: (a) from the GVEP in topological vector spaces to the GVQEP in G-convex spaces without linear structure; (b) the condition (6) is weaker the the condition (iv)(c) of Theorem 2.1 in [14].

Theorem 4.4. *Let $(X;\Gamma)$ be a G-convex space, K be a non-empty compact subset of X and Y be a nonempty set. Let $F, G : X \times X \to 2^Y$, $A : X \to 2^X$ and $C, D : X \to 2^Y$ be set-valued mappings. Suppose that the following conditions are satisfied:*

(1) $A : X \to 2^X$ has nonempty closed G-convex values,

(2) $A^{-1} : X \to 2^X$ is compactly open-valued,

(3) for all $x \in X$, $F(x,x) \not\subseteq C(x)$,

(4) for all $x, y \in X$, $F(x,y) \not\subseteq C(x)$ implies $G(x,y) \subseteq D(y)$,

(5) for each $y \in X$, the set $\{x \in X : G(x,y) \subseteq D(y)\}$ is compactly closed in X,

(6) for each $x \in X$, the set $\{y \in X : F(x,y) \subseteq C(x)\}$ is G-convex,

(7) for each $x \in A(x)$, $G(x,y) \subseteq D(y)$ for all $y \in A(x)$ implies $F(x,y) \not\subseteq C(x)$ for all $y \in A(x)$,

(8) for each $N \in \mathcal{F}(X)$, there exists a compact G-convex subset L_N of X containing N such that for each $y \in L_N \setminus K$, there is an $x \in L_N$ satisfying $y \in (G\text{-co}(A \cap P))^{-1}(x)$ where $P(x) = \{y \in X : G(x,y) \not\subseteq D(y)\}$ for each $x \in X$.

Then there exists an $\hat{x} \in K$ such that

$$\hat{x} \in A(\hat{x}) \quad \text{and} \quad F(\hat{x}, y) \not\subseteq C(\hat{x}), \quad y \in A(\hat{x}).$$

Proof. By (1) and (2), A has nonempty closed G-convex values and for each $y \in X$, $A^{-1} : X \to 2^X$ is compactly open-valued, then the conditions (1) and (2) of Corollary 3.1 with $A = B$ are satisfied. Define two set-valued mappings $P, Q : X \to 2^X$ by

$$P(x) = \{y \in X : G(x, y) \not\subseteq D(y)\}, \quad Q(x) = \{y \in X : F(x, y) \subseteq C(x)\}.$$

It follows from the condition (5) that, for each $y \in X$,

$$\begin{aligned} P^{-1}(y) &= \{x \in X : y \in P(x)\} = \{x \in X : G(x, y) \not\subseteq D(y)\} \\ &= X \setminus \{x \in X : G(x, y) \subseteq D(y)\} \end{aligned}$$

is compactly open in X and hence the condition (3) of Corollary 3.1 is satisfied. The condition (3) implies that for each $x \in X$, $x \notin Q(x)$. The condition (4) implies that $P(x) \subseteq Q(x)$ for each $x \in X$. Note that $Q(x)$ is G-convex by (6), we have $G\text{-co}\, P(x) \subset Q(x)$ for each $x \in X$. It follows that for each $x \in X$, $x \notin G\text{-co}\, P(x)$ and the condition (4) of Corollary 3.1 is satisfied. The condition (8) implies that the condition (5) of Corollary 3.1 holds. By Corollary 3.1 with $A = B$, there exists $\hat{x} \in K$ such that $\hat{x} \in A(\hat{x})$ and $A(\hat{x}) \bigcap P(\hat{x}) = \emptyset$. Hence we have that $\hat{x} \in A(\hat{x})$ and $G(\hat{x}, y) \subset D(y)$ for each $y \in A(\hat{x})$. It follows from the condition (7) that $F(\hat{x}, y) \not\subseteq C(\hat{x})$ for each $y \in A(\hat{x})$. This completes the proof.

Remark 4.4. If $A(x) = X$ for each $x \in X$, then Theorem 4.4 reduces to Theorem 4.3 of Ding-Park [13]. Hence Theorem 4.4 generalizes Theorem 4.3 of Ding-Park [13] from the GVEP to the GVQEP. If X is compact, by letting $K = X = L_N$ for each $N \in \mathcal{F}(X)$, then the condition (8) is satisfied trivially. Hence Theorem 4.3 further generalizes Theorem 1 of Oettli and Schläger [28] in follows ways: (a) from the GVEP in topological vector spaces to the GVQEP in G-convex spaces without linear structure; (b) X may be noncompact G-convex space without linear structure; (c) Y may be any nonempty set without topology. It is clear that our argument methods in Theorems 4.1-4.4 are quite different from that in [1], [2], [6], [7], [9]-[11], [13], [14], [21]-[28], [32].

Theorem 4.5. *Let $(X; \Gamma)$ be a G-convex space, K be a non-empty compact subset of X and Y be a topological vector space. Let $F : X \times X \to 2^Y$, $A : X \to 2^X$ and $P : X \to 2^Y$ be set-valued mappings such that for each $x \in X$, $P(x)$ is a closed convex cone with $P(x) \neq Y$ and $\operatorname{int} P(x) \neq \emptyset$. Suppose the following conditions are satisfied:*

(1) *$A : X \to 2^X$ has nonempty closed G-convex values,*

(2) *$A^{-1} : X \to 2^X$ is compactly open-valued,*

(3) *for all $x \in X$, $F(x, x) \not\subseteq -\operatorname{int} P(x)$,*

(4) *for all $x, y \in X$, $F(x, y) \not\subseteq -\operatorname{int} P(x)$ implies $F(y, x) \subseteq -P(y)$,*

(5) *for each $y \in X$ the set $\{x \in X : F(y, x) \subseteq -P(y)\}$ is compactly open in X,*

(6) *$F(x, y)$ is G-P_x-quasi-convex in y,*

(7) *for each $x \in A(x)$, $F(y, x) \subseteq -P(y)$ for all $y \in A(x)$ implies $F(x, y) \not\subseteq -\operatorname{int} P(x)$ for all $y \in A(x)$,*

(8) *for each $N \in \mathcal{F}(X)$, there exists a compact G-convex subset L_N of X containing N such that for each $y \in L_N \setminus K$, there is an $x \in L_N$ satisfying $y \in (G\text{-co}(A \cap H))^{-1}(x)$ where $H(x) = \{y \in X : F(y, x) \subseteq -P(y)\}$ for each $x \in X$.*

Then there exists an $\hat{x} \in K$ such that

$$\hat{x} \in A(\hat{x}) \quad \text{and} \quad F(\hat{x}, y) \not\subseteq -\operatorname{int} P(\hat{x}), \quad y \in A(\hat{x}).$$

Proof. For each $x, y \in X$ Let $G(x, y) = F(y, x)$, $C(x) = -\operatorname{int} P(x)$, and $D(x) = -P(x)$. We prove that for each $x \in X$, the set $Q(x) = \{y \in X : F(x, y) \subseteq -\operatorname{int} P(x)\}$ is G-convex.

For any $N = \{y_1, \cdots, y_n\} \in \mathcal{F}(Q(x))$ and any $y \in \Gamma(N)$, by the condition (6), there exist $\alpha_i \geq 0$ for $i = 1, \cdots, n$ with $\sum_{i=1}^{n} \alpha_i = 1$ such that

$$F(x,y) \subseteq \sum_{i=1}^{n} \alpha_i F(x, y_i) - P(x) \subseteq -\operatorname{int} P(x) - P(x) = -\operatorname{int} P(x).$$

It follows that $\Gamma(N) \subseteq Q(x)$. Hence for each $x \in X$, the set $\{y \in X : F(x,y) \subseteq -\operatorname{int} P(x)\}$ is G-convex. All conditions of Theorem 4.3 are satisfied. The conclusion of Theorem 4.5 holds from Theorem 4.3. This completes the proof.

References

1. Q. H. Ansari, W. Oetttli and D. Schläger, *A generalization of vectorial equilibria*, Math. Methods Opera. Res. **46** (1997), 147–152.
2. Q. H. Ansari and J. C. Yao, *An existence result for the generalized vector equilibrium problem*, Appl. Math. Lett. **12(8)** (1999), 53–56.
3. J. P. Aubin and A. Cellina, *Differential inclusions*, Springer Verlag, Berlin, Germany, 1994.
4. C. Bardaro and R. Ceppitelli, *Applications of generalized KKM theorem to variational inequalities*, J. Math. Anal. Appl. **137** (1989), 46–58.
5. K. C. Borglin and H. Keiding, *Existence of equilibrium actions of equilibrium*, J. Math. Econom. **3** (1976), 313–316.
6. M. P. Chen, L. J. Lin and S. Park, *Remarks on generalized quasi-equilibrium problems*, Nonlinear Anal. **52(2)** (2003), 433–444.
7. M. S. R. Chowdhury, E. Tarafdar and K. K. Tan, *Minimax inequalities on G-convex spaces with applications to generalized games*, Nonlinear Anal. **43** (2001), 253–275.
8. X. P. Ding, *Fixed points, minimax inequalities and equilibria of noncompact abstract economies*, Taiwanese J. Math. **2(1)** (1998), 25–55.
9. ———, *Quasi-equilibrium problems with applications to infinite optimization and constrained games in general topological spaces*, Appl. Math. Lett. **13** (2000), 21–26.
10. ———, *Quasi-equilibrium problems and constrained multi-objective games in generalized convex spaces*, Appl. Math. Mech. **22(2)** (2001), 160–172.
11. ———, *Maximal element principles on generalized convex spaces and their applications*, in: Mathematical Analysis and Applications, Vol. 4, Taylor & Francis, London, 2002, pp. 149–174.
12. ———, *Generalized G-KKM theorems in generalized convex space and their applications*, J. Math. Anal. Appl. **266** (2002), 21–37.
13. X. P. Ding and J. Y. Park, *Fixed points and generalized vector equilibria in G-convex spaces*, Indian J. Pure Appl. **34(6)** (2003), 973–990.
14. ———, *Generalized vector equilibrium problems in generalized convex spaces*, J. Optim. Theory Appl. **120(2)** (2004), in press.
15. X. P. Ding and K. K. Tan, *Fixed point theorems and equilibria of noncompact generalized games*, Fixed Point Theory and Applications (K. K. Tan, eds.), World Scientific, Singapore, 1992, pp. 80–96.
16. ———, *On equilibria of non-compact generalized games*, J. Math. Anal. Appl. **177** (1993), 226–238.
17. X. P. Ding and E. Tarafdar, *Fixed point theorems and existence of equilibrium points of noncompact abstract economies*, Nonlinear World **1** (1994), 319–340.
18. F. Giannessi, *Theorems of alternative, quadratic programs and complementarity problems*, Variational Inequalities and Complementarity Problems (R. W. Cottle, F. Giannessi and J. L. Lions, Eds.), J. Wiley Sons, New York, 1980, pp. 151–165.
19. ———, *Vector variational inequalities and vector equilibria*, Kluwer Academic Publishers, London, 2000.
20. C. D. Horvath, *Contractibility and generalized convexity*, J. Math. Anal. Appl. **156** (1991), 341–357.
21. L. J. Lin and S. Park, *On some quasi-equilibrium problems*, J. Math. Anal. Appl. **224** (1998), 167–181.
22. L. J. Lin and Z. T. Yu, *Fixed point theorems and equilibrium problems*, Nonlinear Anal. **43** (2001), 987–999.
23. L. J. Lin, Z. T. Yu and G. Kassay, *Existence of equilibria for multivalued Mappings and its application to vectorial equilibria*, J. Optim. Theory Appl. **114(1)** (2002), 189–208.
24. L. J. Lin and Z. T. Yu, *On some equilibrium problems for multimaps*, J. Comput. Appl. Math. **129** (2001), 171–183.
25. W. Oettli and D. Schlager, *Generalized equilibria and generalized monotonicity*, Functional Analysis with Current Applications in Science, Technology and Industry (M. Brokate and A. H. Siddiqi, Eds.), Longman, London, England, 1998, pp. 145–154.

26. ———, *Generalized vector equilibria and generalized monotonicity*, In Functional Analysis with Current Applications in Science, Technology and Industry (M. Brokate and A. H. Sddiqi, Eds.); Pitman Research Notes in Mathematics, Series No. 77, Longman, 1998, pp. 145–154.
27. ———, *Existence of equilibria for monotone mulivalued mappings*, Math. Methods Opera. Res. **48** (1998), 219–228.
28. ———, *Existence of equilibria for g-monotone mappings*, In Nonlinear Analysis and Convex Analysis (W. Takahashi and T. Tanaka, Eds.), World Scientific Pub., Singapore, 1999, pp. 26–33.
29. S. Park, *Fixed points of better admissible maps on generalized convex spaces*, J. Korean Math. Soc. **37(6)** (2000), 885–899.
30. S. Park and H. Kim, *Coincidence theorems for admissible multifunctions on generalized convex spaces*, J. Math. Anal. Appl. **197(1)** (1996), 173–187.
31. ———, *Foundations of the KKM theory on generalized convex spaces*, J. Math. Anal. Appl. **209(2)** (1997), 551–571.
32. W. Song, *Vector equilibrium problems with set-valued mappings*, In Vector Variational Inequalities and Vector Equilibria (F. Giannessi, Eds.), Kluwer Academic Publishers, London, 2000, pp. 403–422.
33. K. K. Tan and X. L. Zhang, *Fixed point theorems on G-convex spaces and applications*, Nonlinear Funct. Anal. Appl. **1** (1996), 1–19.
34. K. K. Tan and X. Z. Yuan, *A minimax inequality with applications to existence of equilibrium points*, Bull. Austral. Math. Soc. **47** (1993), 483–503.
35. S. Toussaint, *On the existence of equilibria in economies with infinitely many commoddities and without ordered preferences*, J. Econom. Theory **33** (1984), 98–115.
36. C. I. Tulcea, *On the approximation of upper semicontinuous correspondences and equilibriums of generalized games*, J. Math. Anal. Appl. **136** (1988), 267–289.
37. N. C. Yannelis, *Maximal elements over non-compact subsets of linear topological spaces*, Econom. Lett. **17** (1985), 133–136.
38. N. C. Yannelis and N. D. Prabhakar, *Existence of maxiaml elements and equilibria in linear topological spaces*, J. Math. Econom. **12** (1983), 133–145.
39. X. Z. Yuan, *The study of equilibria for abstract economics in topological vector spaces–a unified approach*, Nonlinear Anal. **37** (1999), 409–430.

In: Inequality Theory and Applications, Volume 4
Editors: Yeol Je Cho et al., pp. 61–64

ISBN 1-59454-874-9

A GENERALISATION OF AN OSTROWSKI INEQUALITY IN INNER PRODUCT SPACES

SEVER S. DRAGOMIR[1] AND ANCA C. GOŞA[2]

ABSTRACT. A generalisation of inner product spaces of an inequality due to Ostrowski and applications for sequences and integrals are given.

1. Introduction

In 1951, A. M. Ostrowski [2, p. 289] obtained the following result (see also [2, p. 92]).

Theorem 1. *Suppose that* $\mathbf{a} = (a_1, \dots, a_n)$, $\mathbf{b} = (b_1, \dots, b_n)$ *and* $\mathbf{x} = (x_1, \dots, x_n)$ *are real* $n-$*tuples such that* $\mathbf{a} \neq 0$ *and*

$$\sum_{i=1}^{n} a_i x_i = 0 \quad \text{and} \quad \sum_{i=1}^{n} b_i x_i = 1. \tag{1.1}$$

Then

$$\sum_{i=1}^{n} x_i^2 \geq \frac{\sum_{i=1}^{n} a_i^2}{\sum_{i=1}^{n} a_i^2 \sum_{i=1}^{n} b_i^2 - \left(\sum_{i=1}^{n} a_i b_i\right)^2} \tag{1.2}$$

with the equality if and only if

$$x_k = \frac{b_k \sum_{i=1}^{n} a_i^2 - a_k \sum_{i=1}^{n} a_i b_i}{\sum_{i=1}^{n} a_i^2 \sum_{i=1}^{n} b_i^2 - \left(\sum_{i=1}^{n} a_i b_i\right)^2}, \quad k = 1, \cdots, n. \tag{1.3}$$

Another similar result due to Ostrowski which is far less known and obtained in the same work [2, p. 290] (see also [1, p. 94]), is the following one.

Received June 12, 2004.
2000 *Mathematics Subject Classification.* 26D15, 46C99.
Key words and phrases. Ostrowski's inequality, Inner products.

[1] School of Computer Science and Mathematics, Victoria University of Technology, PO Box 14428, MCMC 8001, Victoria, Australia. Email: sever@matilda.vu.edu.au
[2] College No. 12, Reşiţa, Jud. Caraş-Severin, R0-1700, Reşiţa, Romania. E-mail: ancagosa@hotmail.com

Theorem 2. *Let* **a**, **b** *and* **x** *be* $n-$*tuples of real numbers with* $\mathbf{a} \neq 0$ *and*

$$\sum_{i=1}^{n} a_i x_i = 0 \quad \text{and} \quad \sum_{i=1}^{n} x_i^2 = 1. \tag{1.4}$$

Then

$$\frac{\sum_{i=1}^{n} a_i^2 \sum_{i=1}^{n} b_i^2 - \left(\sum_{i=1}^{n} a_i b_i\right)^2}{\sum_{i=1}^{n} a_i^2} \geq \left(\sum_{i=1}^{n} b_i x_i\right)^2. \tag{1.5}$$

If **a** *and* **b** *are not proportional, then the equality holds in* (1.5) *if and only if*

$$x_k = q \cdot \frac{b_k \sum_{i=1}^{n} a_i^2 - a_k \sum_{i=1}^{n} a_i b_i}{\left(\sum_{k=1}^{n} a_k^2\right)^{\frac{1}{2}} \left[\sum_{i=1}^{n} a_i^2 \sum_{i=1}^{n} b_i^2 - \left(\sum_{i=1}^{n} a_i b_i\right)^2\right]^{\frac{1}{2}}}, \quad k \in \{1, \ldots, n\}, \tag{1.6}$$

with $q \in \{-1, 1, \}$.

The case of equality which was neither mentioned in [1] nor in [2] is considered in Remark 1.

In the present paper, by the use of an elementary argument based on Schwarz's inequality, a natural generalisation in inner-product spaces of (1.5) is given. The case of equality is analyzed. Applications for sequences and integrals are also provided.

2. The Main Results

The following theorem holds.

Theorem 3. *Let* $(H, \langle \cdot, \cdot \rangle)$ *be a real or complex inner product space and* $a, b \in H$ *be two linearly independent vectors. If* $x \in H$ *is such that*

$$\langle x, a \rangle = 0 \quad \text{and} \quad \|x\| = 1, \tag{i}$$

then

$$\frac{\|a\|^2 \|b\|^2 - |\langle a, b \rangle|^2}{\|a\|^2} \geq |\langle x, b \rangle|^2. \tag{2.1}$$

The equality holds in (2.1) *if and only if*

$$x = \nu \left(b - \frac{\overline{\langle a, b \rangle}}{\|a\|^2} \cdot a \right), \tag{2.2}$$

where $\nu \in \mathbb{K}$ $(\mathbb{C}, \mathbb{R})$ *is such that*

$$|\nu| = \frac{\|a\|}{\left[\|a\|^2 \|b\|^2 - |\langle a, b \rangle|^2\right]^{\frac{1}{2}}}. \tag{2.3}$$

Proof. We use Schwarz's inequality in the inner product space H, i.e.,

$$\|u\|^2 \|v\|^2 \geq |\langle u, v \rangle|^2, \quad u, v \in H, \tag{2.4}$$

with the equality if and only if there is a scalar $\alpha \in \mathbb{K}$ such that

$$u = \alpha v. \tag{2.5}$$

If we apply (2.4) for $u = z - \frac{\langle z,c\rangle}{\|c\|^2} \cdot c$, $v = d - \frac{\langle d,c\rangle}{\|c\|^2} \cdot c$, where $c \neq 0$ and $c, d, z \in H$, and taking into account that

$$\left\| z - \frac{\langle z, c\rangle}{\|c\|^2} \cdot c \right\|^2 = \frac{\|z\|^2 \|c\|^2 - |\langle z, c\rangle|^2}{\|c\|^2},$$
$$\left\| d - \frac{\langle d, c\rangle}{\|c\|^2} \cdot c \right\|^2 = \frac{\|d\|^2 \|c\|^2 - |\langle d, c\rangle|^2}{\|c\|^2}$$

and

$$\left\langle z - \frac{\langle z, c\rangle}{\|c\|^2} \cdot c, d - \frac{\langle d, c\rangle}{\|c\|^2} \cdot c \right\rangle = \frac{\langle z, d\rangle \|c\|^2 - \langle z, c\rangle \langle c, d\rangle}{\|c\|^2},$$

then we deduce the inequality

$$\left[\|z\|^2 \|c\|^2 - |\langle z, c\rangle|^2 \right] \left[\|d\|^2 \|c\|^2 - |\langle d, c\rangle|^2 \right] \geq \left| \langle z, d\rangle \|c\|^2 - \langle z, c\rangle \langle c, d\rangle \right|^2 \tag{2.6}$$

with the equality if and only if there is a $\beta \in \mathbb{K}$ such that

$$z = \frac{\langle z, c\rangle}{\|c\|^2} \cdot c + \beta \left(d - \frac{\langle d, c\rangle}{\|c\|^2} \cdot c \right). \tag{2.7}$$

If in (2.6) we choose $z = x$, $c = a$ and $d = b$, where a and x satisfy (i), then we deduce

$$\|a\|^2 \left[\|a\|^2 \|b\|^2 - |\langle a, b\rangle|^2 \right] \geq \left[\langle x, b\rangle \|a\|^2 \right]^2,$$

which is clearly equivalent to (2.1).

The equality holds in (2.1) if and only if

$$x = \nu \left(b - \frac{\overline{\langle a, b\rangle}}{\|a\|^2} \cdot a \right),$$

where $\nu \in \mathbb{K}$ satisfies the condition

$$1 = \|x\| = |\nu| \left\| b - \frac{\overline{\langle a, b\rangle}}{\|a\|^2} \cdot a \right\| = |\nu| \left[\frac{\|a\|^2 \|b\|^2 - |\langle a, b\rangle|^2}{\|a\|^2} \right]^{\frac{1}{2}} \tag{2.8}$$

and the theorem is thus proved.

The following particular cases hold.

1. If $\mathbf{a}, \mathbf{b}, \mathbf{x} \in \ell^2(\mathbb{K})$, $\mathbb{K} = \mathbb{C}, \mathbb{R}$, where

$$\ell^2(\mathbb{K}) := \left\{ x = (x_i)_{i \in \mathbb{N}}, \sum_{i=1}^{\infty} |x_i|^2 < \infty \right\}$$

with $\mathbf{a}, \mathbf{b}$ linearly independent and

$$\sum_{i=1}^{\infty} x_i \overline{a_i} = 0, \quad \sum_{i=1}^{\infty} |x_i|^2 = 1, \tag{a}$$

then

$$\frac{\sum_{i=1}^{\infty}|a_i|^2\sum_{i=1}^{\infty}|b_i|^2-\left|\sum_{i=1}^{\infty}a_i\overline{b_i}\right|^2}{\sum_{i=1}^{\infty}|a_i|^2}\geq\left|\sum_{i=1}^{\infty}x_i\overline{b_i}\right|^2. \tag{2.9}$$

The equality holds in (2.9) if and only if

$$x_i=\nu\left[b_i-\frac{\sum_{k=1}^{\infty}a_k\overline{b_k}}{\sum_{k=1}^{\infty}|a_k|^2}\cdot a_i\right],\quad i\in\{1,2,\dots\}, \tag{2.10}$$

with $\nu\in\mathbb{K}$ is such that

$$|\nu|=\frac{\left(\sum_{k=1}^{\infty}|a_k|^2\right)^{\frac{1}{2}}}{\left[\sum_{k=1}^{\infty}|a_k|^2\sum_{k=1}^{\infty}|b_k|^2-\left|\sum_{k=1}^{\infty}a_k\overline{b_k}\right|^2\right]^{\frac{1}{2}}}. \tag{2.11}$$

Remark 1. The case of equality in (1.5) is obviously a particular case of the above. We omit the details.

Remark 2. If $f,g,h\in L^2(\Omega,m)$, where Ω is an $m-$measurable space and

$$L^2(\Omega,m):=\left\{f:\Omega\to\mathbb{K},\ \int_\Omega|f(x)|^2\,dm(x)<\infty\right\}$$

with f,g being linearly independent and

$$\int_\Omega h(x)\overline{f(x)}dm(x)=0,\quad \int_\Omega|h(x)|^2\,dm(x)=1, \tag{2.12}$$

then

$$\begin{aligned}&\frac{\int_\Omega|f(x)|^2\,dm(x)\int_\Omega|g(x)|^2\,dm(x)-\left|\int_\Omega f(x)\overline{g(x)}dm(x)\right|^2}{\int_\Omega|f(x)|^2\,dm(x)}\\&\geq\left|\int_\Omega h(x)\overline{g(x)}dm(x)\right|^2.\end{aligned} \tag{2.13}$$

The equality holds in (2.13) if and only if

$$h(x)=\nu\left[g(x)-\frac{\int_\Omega g(x)\overline{f(x)}dm(x)}{\int_\Omega|f(x)|^2\,dm(x)}f(x)\right]\quad\text{for a.e. } x\in\Omega$$

and $\nu\in\mathbb{K}$ with

$$|\nu|=\frac{\left(\int_\Omega|f(x)|^2\,dm(x)\right)^{\frac{1}{2}}}{\left[\int_\Omega|f(x)|^2\,dm(x)\int_\Omega|g(x)|^2\,dm(x)-\left|\int_\Omega f(x)\overline{g(x)}dm(x)\right|^2\right]^{\frac{1}{2}}}.$$

References

1. D. S. Mitrinović, J. E. Pečarić and A. M. Fink, *Classical and New Inequalities in Analysis*, Kluwer Academic Publishers, Dordrecht/Boston/London, 1993.
2. A. M. Ostrowski, *Varlesungen über Differential und Integralrechnung II*, Birkhäuser, Basel, 1951.

In: Inequality Theory and Applications, Volume 4
Editors: Yeol Je Cho et al., pp. 65–69

ISBN 1-59454-874-9

ON THE REMAINDER ESTIMATE IN THE GENERALISED TAYLOR FORMULA

S. S. DRAGOMIR[1] AND C. I. PREDA[2]

ABSTRACT. In this note, we point out an estimate for the remainder in the generalised Taylor formula which improves the recent result by Matić, Pečarić and Ujević [2].

1. Introduction

In the recent paper [2], M. Matić, J. E. Pečarić and N. Ujević proved the following generalised Taylor formula.

Theorem 1. *Let $\{P_n\}_{n\in\mathbb{N}}$ be a harmonic sequence of polynomials, that is, $P_n'(t) = P_{n-1}(t)$ for $n \geq 1$, $n \in \mathbb{N}$, $P_0(t) = 1$, $t \in \mathbb{R}$. Further, let $I \subset \mathbb{R}$ be a closed interval and $a \in I$. If $f : I \to \mathbb{R}$ is a function such that, for some $n \in \mathbb{N}$, $f^{(n)}$ is absolutely continuous, then*

$$f(x) = \tilde{T}_n(f;a,x) + \tilde{R}_n(f;a,x), \quad x \in I, \tag{1.1}$$

where

$$\tilde{T}_n(f;a,x) = f(a) + \sum_{k=1}^{n} (-1)^{k+1} \left[P_k(x) f^{(k)}(x) - P_k(a) f^{(k)}(a)\right] \tag{1.2}$$

and

$$\tilde{R}_n(f;a,x) = (-1)^n \int_a^x P_n(t) f^{(n+1)}(t)dt. \tag{1.3}$$

For some particular instances of harmonic sequences, they obtained the following Taylor-like expansions:

$$f(x) = T_n^{(M)}(f;a,x) + R_n^{(M)}(f;a,x), \quad x \in I, \tag{1.4}$$

Received June 12, 2004.
2000 *Mathematics Subject Classification.* Primary 26D15; Secondary 41A58.
Key words and phrases. Taylor expansion, Analytic Inequalities, Cauchy-Buniakowski-Schwartz inequality.
[1] School of Computer Science & Mathematics, Victoria University of Technology, PO Box 14428, Melbourne City MC, Victoria 8001, Australia. E-mail: sever.dragomir@vu.edu.au
[2] Department of Mathematics, West University of Timisoara, B-dul V. Parvan, No. 4, Timisoara RO-1900, Romania. E-mail: preda@math.uvt.ro (http://rgmia.vu.edu.au/SSDra gomirWeb.html)

where

$$T_n^{(M)}(f;a,x) = f(a) + \sum_{k=1}^{n} \frac{(x-a)^k}{2^k k!}\left[f^{(k)}(a) + (-1)^{k+1} f^{(k)}(x)\right], \tag{1.5}$$

$$R_n^{(M)}(f;a,x) = \frac{(-1)^n}{n!}\int_a^x \left(t - \frac{a+x}{2}\right)^n f^{(n+1)}(t)dt \tag{1.6}$$

and

$$f(x) = T_n^{(B)}(f;a,x) + R_n^{(B)}(f;a,x), \quad x \in I, \tag{1.7}$$

where

$$\begin{aligned} T_n^{(B)}(f;a,x) &= f(a) + \frac{x-a}{2}[f'(x) + f'(a)] \\ &- \sum_{k=1}^{[\frac{n}{2}]} \frac{(x-a)^{2k}}{(2k)!} B_{2k}\left[f^{(2k)}(x) - f^{(2k)}(a)\right], \end{aligned} \tag{1.8}$$

and $[r]$ is the integer part of r. Here B_{2k} are the Bernoulli numbers, and

$$R_n^{(B)}(f;a,x) = (-1)^n \frac{(x-a)^n}{n!}\int_a^x B_n\left(\frac{t-a}{x-a}\right) f^{(n+1)}(t)dt, \tag{1.9}$$

where $B_n(\cdot)$ are the Bernoulli polynomials, respectively. In addition, they proved that

$$f(x) = T_n^{(E)}(f;a,x) + R_n^{(E)}(f;a,x), \quad x \in I, \tag{1.10}$$

where

$$\begin{aligned} &T_n^{(E)}(f;a,x) \\ &= f(a) + 2\sum_{k=1}^{[\frac{n+1}{2}]} \frac{(x-a)^{2k-1}(4^k-1)}{(2k)!} B_{2k}\left[f^{(2k-1)}(x) + f^{(2k-1)}(a)\right] \end{aligned} \tag{1.11}$$

and

$$R_n^{(E)}(f;a,x) = (-1)^n \frac{(x-a)^n}{n!}\int_a^x E_n\left(\frac{t-a}{x-a}\right) f^{(n+1)}(t)dt, \tag{1.12}$$

where $E_n(\cdot)$ are the Euler polynomials.

In [1], S. S. Dragomir was the first author to introduce the perturbed Taylor formula

$$f(x) = T_n(f;a,x) + \frac{(x-a)^{n+1}}{(n+1)!}\left[f^{(n)};a,x\right] + G_n(f;a,x), \tag{1.13}$$

where

$$T_n(f;a,x) = \sum_{k=0}^{n} \frac{(x-a)^k}{k!} f^{(k)}(a) \tag{1.14}$$

and

$$[f^{(n)};a,x] := \frac{f^{(k)}(x) - f^{(k)}(a)}{x-a}$$

and had the idea to estimate the remainder $G_n(f;a,x)$ by using Grüss and Chebychev type inequalities.

In [2], the authors generalised and improved the results from [1]. We mention here the following result obtained via a pre-Grüss inequality (see [2, Theorem 3]).

Theorem 2. *Let $\{P_n\}_{n\in\mathbb{N}}$ be a harmonic sequence of polynomials. Let $I \subset \mathbb{R}$ be a closed interval and $a \in I$. Suppose $f : I \to \mathbb{R}$ is as in Theorem 1. Then, for all $x \in I$, we have the perturbed generalised Taylor formula:*

$$\begin{aligned} f(x) = \tilde{T}_n(f; a, x) + (-1)^n [P_{n+1}(x) - P_{n+1}(a)] \left[f^{(n)}; a, x\right] \\ + \tilde{G}_n(f; a, x). \end{aligned} \tag{1.15}$$

For $x \geq a$, the remainder $\tilde{G}(f; a, x)$ satisfies the estimate

$$\left|\tilde{G}_n(f; a, x)\right| \leq \frac{x-a}{2}\sqrt{T(P_n, P_n)}[\Gamma(x) - \gamma(x)], \tag{1.16}$$

provided that $f^{(n+1)}$ is bounded and

$$\Gamma(x) := \sup_{t\in[a,x]} f^{(n+1)}(t) < \infty, \quad \gamma(x) := \inf_{t\in[a,x]} f^{(n+1)}(t) > -\infty, \tag{1.17}$$

where $T(\cdot, \cdot)$ is the Chebychev functional on the interval $[a, x]$, that is, we recall

$$T(g, h) := \frac{1}{x-a}\int_a^x g(t)h(t)dt - \frac{1}{x-a}\int_a^x g(t)dt \cdot \frac{1}{x-a}\int_a^x h(t)dt. \tag{1.18}$$

The main aim of the present note is to improve the inequality (1.16) as follows.

2. The Main Results

The following result holds.

Theorem 3. *Assume that $\{P_n\}_{n\in\mathbb{N}}$ is a sequence of harmonic polynomials and $f : I \to \mathbb{R}$ is such that $f^{(n)}$ is absolutely continuous and $f^{(n+1)} \in L_2(I)$. If $x \geq a$, then we have the inequality*

$$\begin{aligned} &\left|\tilde{G}_n(f; a, x)\right| \\ &\leq (x-a)[T(P_n, P_n)]^{\frac{1}{2}} \left[\frac{1}{x-a}\|f^{(n+1)}\|_2^2 - \left(\left[f^{(n)}; a, x\right]\right)^2\right]^{\frac{1}{2}} \\ &\left(\leq \frac{x-a}{2}[T(P_n, P_n)]^{\frac{1}{2}}[\Gamma(x) - \gamma(x)], \quad \text{if } f^{(n+1)} \in L_\infty[a, x]\right), \end{aligned} \tag{2.1}$$

where $\|\cdot\|_2$ is the usual Euclidean norm on $[a, x]$, i.e.,

$$\|f^{(n+1)}\|_2 = \left(\int_a^x \left|f^{(n+1)}(t)\right|^2 dt\right)^{\frac{1}{2}}.$$

Proof. Recall Korkine's identity for the mappings h, g, which can be easily proved by direct computation:

$$T(h, g) := \frac{1}{2(x-a)^2}\int_a^x\int_a^x (h(t) - h(s))(g(t) - g(s))dtds. \tag{2.2}$$

Using (2.2), we have

$$\begin{aligned} &(-1)^n \int_a^x P_n(t) f^{(n+1)}(t)dt - \frac{(-1)^n}{x-a}\int_a^x f^{(n+1)}(t)dt \cdot \int_a^x P_n(t)dt \\ &= \frac{(-1)^n}{2}\cdot\frac{1}{x-a}\int_a^x\int_a^x (P_n(t) - P_n(s))\left(f^{(n+1)}(t) - f^{(n+1)}(s)\right)dtds \end{aligned} \tag{2.3}$$

and then, by the equalities (1.1)and (1.15), we have the following representation of the remainder in the perturbed formula (1.15)

$$\tilde{G}_n(f;a,x) = \frac{(-1)^n}{2}\cdot\frac{1}{x-a}\int_a^x\int_a^x (P_n(t)-P_n(s))\big(f^{(n+1)}(t)-f^{(n+1)}(s)\big)dtds, \tag{2.4}$$

which is an identity that is interesting in itself as well. Using now the Cauchy-Buniakowsky-Schwartz integral inequality for double integrals, we have

$$\begin{aligned}
&|\tilde{G}_n(f;a,x)| \\
&\leq \frac{1}{2(x-a)}\left[\int_a^x\int_a^x (P_n(t)-P_n(s))^2 dtds\right]^{\frac{1}{2}} \\
&\quad\times\left[\int_a^x\int_a^x \big(f^{(n+1)}(t)-f^{(n+1)}(s)\big)^2 dtds\right]^{\frac{1}{2}} \\
&= (x-a)[T(P_n,P_n)]^{\frac{1}{2}}\left[\frac{1}{x-a}\|f^{(n+1)}\|_2^2-\left(\frac{1}{x-a}\int_a^x f^{(n+1)}(t)dt\right)^2\right]^{\frac{1}{2}}
\end{aligned}$$

and the first inequality in (2.1)is proved.

The second inequality is obvious by the Grüss inequality

$$\begin{aligned}
&\frac{1}{x-a}\int_a^x \left[f^{(n+1)}(t)\right]^2 dt-\left(\frac{1}{x-a}\int_a^x f^{(n+1)}(t)dt\right)^2 \\
&\leq \frac{1}{4}[\Gamma(x)-\gamma(x)]^2
\end{aligned} \tag{2.5}$$

and the theorem is proved.

Remark 1. If $f^{(n+1)}$ is unbounded on (a,x) but $f^{(n+1)}\in L_2(a,x)$, then the first inequality in (2.1)can still be applied, but not the Matić-Pečarić-Ujević result (1.16) which requires the boundedness of the derivative $f^{(n+1)}$.

The following corollary improves Corollary 3 of [2], which deals with the estimation of the remainder for the particular perturbed Taylor-like formulae (1.4), (1.7) and (1.10).

Corollary 1. *With the assumptions in Theorem 3, we have the following inequalities*

$$\left|\tilde{G}_n^{(M)}(f;a,x)\right| \leq \frac{(x-a)^{n+1}}{n!2^n\sqrt{2n+1}}\times\sigma\big(f^{(n+1)};a,x\big), \tag{2.6}$$

$$\left|\tilde{G}_n^{(B)}(f;a,x)\right| \leq (x-a)^{n+1}\left[\frac{|B_{2n}|}{(2n)!}\right]^{\frac{1}{2}}\times\sigma\big(f^{(n+1)};a,x\big), \tag{2.7}$$

$$\begin{aligned}
&\left|\tilde{G}_n^{(E)}(f;a,x)\right| \\
&\leq 2(x-a)^{n+1}\left[\frac{(4^{n+1}-1)|B_{2n+2}|}{(2n+2)!}-\left[\frac{2(2^{n+2}-1)B_{n+2}}{(n+1)!}\right]^2\right]^{\frac{1}{2}} \\
&\quad\times\sigma\big(f^{(n+1)};a,x\big)
\end{aligned} \tag{2.8}$$

and

$$|G_n(f;a,x)| \leq \frac{n(x-a)^{n+1}}{(n+1)!\sqrt{2n+1}}\times\sigma\big(f^{(n+1)};a,x\big), \tag{2.9}$$

where, as in [2],

$$\tilde{G}_n^{(M)}(f;a,x) = f(x) - T_n^M(f;a,x) - \frac{(x-a)^{n+1}[1+(-1)^n]}{(n+1)!2^{n+1}}\left[f^{(n)};a,x\right];$$

$$\tilde{G}_n^{(B)}(f;a,x) = f(x) - T_n^B(f;a,x);$$

$$\tilde{G}_n^{(E)}(f;a,x) = f(x) - \frac{4(-1)^n(x-a)^{n+1}(2^{n+2}-1)B_{n+2}}{(n+2)!}\left[f^{(n)};a,x\right];$$

$G_n(f;a,x)$ *is as defined by* (1.13),

$$\sigma\left(f^{(n+1)};a,x\right) := \left[\frac{1}{x-a}\|f^{(n+1)}\|_2^2 - \left(\left[f^{(n+1)};a,x\right]\right)^2\right]^{\frac{1}{2}} \tag{2.10}$$

and, $x \geq a,\ f^{(n+1)} \in L_2[a,x]$.

Note that, for all the examples considered in [1] and [2] for f, the quantity $\sigma(f^{(n+1)};a,x)$ can be completely computed and then those particular inequalities may be improved accordingly. We omit the details.

Remark 2. Theorem 3 is an implicit refinement of Theorem 4 from [2] which uses Chebychev's inequality to estimate $\sigma(f^{(n+1)};a,x)$ and of Theorem 5 from [2] which uses Lupaş's inequality to upper bound the same term $\sigma(f^{(n+1)};a,x)$.

References

1. S. S. Dragomir, *New estimation of the remainder in Taylor's formula using Grüss' type inequalities and applications*, Math. Ineq. Appl. **2(2)** (1999), 183–193.
2. M. Matić, J. E. Pečarić and N. Ujević, *On new estimation of the remainder in generalised Taylor's formula*, Math. Ineq. Appl. **2(3)** (1999), 343–361.

In: Inequality Theory and Applications, Volume 4
Editors: Yeol Je Cho et al., pp. 71–84

ISBN 1-59454-874-9

A PERTURBED VERSION OF THE GENERALISED TAYLOR'S FORMULA AND APPLICATIONS

S. S. Dragomir[1] and A. Sofo[1]

Abstract. A Perturbed version of the generalised Taylor's formula for Appell-type polynomials and applications are given.

1. Introduction

In [4], Matić et al. introduced the concept of *harmonic sequences of polynomials* by assuming that the polynomial $\{P_n\}_{n\in\mathbb{N}}$ satisfies the condition

$$P_0 = 1, \;\; P_n'(t) = P_{n-1}(t), \quad t \in \mathbb{R}, \; n \in \mathbb{N}. \tag{1.1}$$

With this assumption, they proved the following generalised Taylor's formula:

Theorem 1. *Let $I \subset \mathbb{R}$ be a closed interval and $a \in I$. If $f : I \to \mathbb{R}$ is any function such that, for some $n \in \mathbb{N}, f^{(n)}$ is absolutely continuous, then, for any $x \in I$,*

$$f(x) = f(a) + \sum_{k=1}^{n} (-1)^{k+1} \left[P_k(x) f^{(k)}(x) - P_k(a) f^{(k)}(a)\right] + R_n(f; a, x), \tag{1.2}$$

where

$$R_n(f; a, x) = (-1)^n \int_a^x P_n(t) f^{(n+1)}(t)dt \tag{1.3}$$

and $\{P_n\}_{n\in\mathbb{N}}$ is a harmonic sequence of polynomials.

As examples of such polynomials, they mentioned the following

$$P_n(t) := \frac{1}{n!}(t - x)^n, \quad t \in \mathbb{R};$$

or

$$P_n(t) := \frac{1}{n!}\Big(t - \frac{a + x}{2}\Big)^n, \quad t \in \mathbb{R};$$

Received June 12, 2004.
2000 *Mathematics Subject Classification.* Primary 26D15; Secondary 26D10.
Key words and phrases. Taylor's formula, Appell polnomials.

[1] School of Computer Science and Mathematics, Victoria University of Technology, P.O. Box 14428, Melbournce City MC, Victoria, 8001, Australia. E-mails: sever@matilda.vu.edu.au (http://rgmia.vu.edu.au/SSDragomirWeb.html) and sofo@matilda.vu.edu.au (http://rgmia.vu.edu.au/sofo)

or

$$P_n(t) := \frac{(x-a)^n}{n!} B_n\Big(\frac{t-a}{x-a}\Big), \quad x \geq 1, \ P_0(t) = 1,$$

where $B_n(\cdot)$ are the *Bernoulli polynomials,* or

$$P_n(t) := \frac{(x-a)^n}{n!} E_n\Big(\frac{t-a}{x-a}\Big), \quad x \geq 1, \ P_0(t) = 1,$$

where $E_n(\cdot)$ are the *Euler polynomials.*

Matić et al. [4] proved the following general estimation result for the remainder $R_n(f; a, x)$.

Corollary 1. *Under the assumptions of Theorem 1 and, if $x \geq a$, then*

$$(1.4) \qquad |R_n(f;a,x)| \leq \begin{cases} \max_{t\in[a,x]} |P_n(t)| \int_a^x |f^{(n+1)}(s)|ds; \\ \int_a^x |P_n(s)|ds \max_{t\in[a,x]} |f^{(n+1)}(t)|; \\ \left(\int_a^x |P_n(s)|^q ds\right)^{\frac{1}{q}} \left(\int_a^x |f^{(n+1)}(s)|^p ds\right)^{\frac{1}{p}}, \\ \qquad \text{where } \frac{1}{p}+\frac{1}{q}=1, \ p>1. \end{cases}$$

Now, if one would choose in (1.2) $f(x) = \int_a^b g(t)dt$ and then put $x = b$, one could state the following generalised integration by parts formula

$$(1.5) \qquad \int_a^b g(t)dt = \sum_{k=1}^n (-1)^{k+1} \left[P_k(b)g^{(k-1)}(b) - P_k(a)g^{(k-1)}(a)\right] + S_n(f;a,b),$$

where

$$(1.6) \qquad S_n(f;a,b) = (-1)^n \int_a^b P_n(t)g^{(n)}(t)dt.$$

Using the classical notation for the Lebesgue norms

$$\|h\|_\infty := \operatorname*{ess\,sup}_{t\in[a,b]} |h(t)|;$$

$$\|h\|_p := \left(\int_a^b |h(t)|^p dt\right)^{\frac{1}{p}}, \quad p \geq 1;$$

the remainder (1.6) may be bounded in the following manner

$$(1.7) \qquad |S_n(g;a,b)| \leq \begin{cases} \|P_n\|_\infty \|g^{(n)}\|_1; \\ \|P_n\|_1 \|g^{(n)}\|_\infty; \\ \|P_n\|_q \|g^{(n)}\|_p, \quad \text{where } \frac{1}{p}+\frac{1}{q}=1, \ p>1. \end{cases}$$

For recent results on Taylor's expansion, see [1] and [3].

2. The Main Results

The following result is a reformulation of Theorem 1 in a more appropriate manner for our further purposes.

Theorem 2. *Assume that the sequence of polynomials $\{P_n(t,x)\}_{x\in\mathbb{N}}$ satisfies the Appell condition [2], i.e.,*

$$\frac{\partial P_n(t,x)}{\partial t} = P_{n-1}(t,x), \quad t,x\in\mathbb{R},\ n\in\mathbb{N} \text{ and } P_0(t,x)=1,\ t,\ x\in\mathbb{R}. \tag{2.1}$$

Let $I\subset\mathbb{R}$ be a closed interval and $a\in I$. If $f: I\to\mathbb{R}$ is any function such that for some $n\in\mathbb{N}, f^{(n)}$ is absolutely continuous, then, for any $x\in I$, we have the representation

$$\begin{aligned} f(x) = f(a) &+ \sum_{k=1}^{n}(-1)^{k+1}\left[P_k(x,x)f^{(k)}(x) - P_k(a,x)f^{(k)}(a)\right] \\ &+ R_n(f;a,x), \end{aligned} \tag{2.2}$$

where

$$R_n(f;a,x) = (-1)^n \int_a^x P_n(t,x) f^{(n+1)}(t)dt. \tag{2.3}$$

Using the notations

$$\|g\|_{[a,b],p} := \left|\int_a^b |g(t)|^p dt\right|^{\frac{1}{p}}, \quad p\geq 1,$$

and

$$\|g\|_{[a,b],\infty} := \underset{\substack{t\in[a,b] \\ (t\in[b,a])}}{\text{ess sup}} |g(t)|,$$

where a,b are arbitrary real numbers, then we can state the following corollary (see [4, Corollary 1])

Corollary 2. *With the assumptions in Theorem 2, we have the bounds for the remainder $R_n(f;a,x)$*

$$|R_n(f;a,x)| \leq \begin{cases} \|P_n(\cdot,x)\|_{[a,x],\infty}\|f^{(n+1)}\|_{[a,x],1}; \\ \|P_n(\cdot,x)\|_{[a,x],q}\|f^{(n+1)}\|_{[a,x],p}, & p>1,\ \frac{1}{p}+\frac{1}{q}=1, \\ & \text{if } f^{(n+1)}\in L_p[a,x]; \\ \|P_n(\cdot,x)\|_{[a,x],1}\|f^{(n+1)}\|_{[a,x],\infty}, & \text{if } f^{(n+1)}\in L_\infty[a,x]. \end{cases} \tag{2.4}$$

The following perturbed version of the above result also holds.

Theorem 3. *Let $P_n(t,x), f, a, x$ be as in Theorem 2. Then, for any $y\in I$, we have the representation*

$$\begin{aligned} f(x) = f(a) &+ \sum_{k=1}^{n}(-1)^{k+1}\left[P_k(x,x)f^{(k)}(x) - P_k(a,x)f^{(k)}(a)\right] \\ &+ (-1)^n\left[P_{n+1}(x,x) - P_{n+1}(a,x)\right]f^{(n+1)}(y) + S_n(f;a,x,y), \end{aligned} \tag{2.5}$$

where the new remainder $S_n(f;a,x,y)$ can be represented in the following manner

$$S_n(f;a,x,y) = (-1)^n \int_a^x P_n(t,x)\left[f^{(n+1)}(t) - f^{(n+1)}(y)\right]dt. \tag{2.6}$$

Proof. We have

$$\begin{aligned}R_n(f;a,x) &= (-1)^n \int_a^x P_n(t,x) f^{(n+1)}(t)dt \\ &= (-1)^n \int_a^x P_n(t,x) f^{(n+1)}(y)dt \\ &\quad + (-1)^n \int_a^x P_n(t,x)\left[f^{(n+1)}(t) - f^{(n+1)}(y)\right]dt \\ &= (-1)^n P_{n+1}(t,x) f^{(n+1)}(y)\Big|_a^x + S_n(f;a,x,y) \\ &= (-1)^n \left[P_{n+1}(x,x) - P_{n+1}(a,x)\right] f^{(n+1)}(y) + S_n(f;a,x,y).\end{aligned}$$

Using (2.2), we deduce (2.5).

Remark 1. Some particular interesting cases are for $y = a$ and $y = x$.

The following corollary for functions f for which $f^{(n+1)}$ is Hölder continuous holds.

Corollary 3. *If the function $f^{(n+1)} : I \to \mathbb{R}$ is of r-H-Hölder type on I, i.e.,*

$$\left|f^{(n+1)}(s) - f^{(n+1)}(u)\right| \leq H|s-u|^r, \quad s,u \in I, \tag{2.7}$$

where $H > 0$ and $r \in [0,1]$ are given, then, for any y between a and x, we have the bound

$$|S_n(f;a,x,y)| \leq \begin{cases} \frac{H}{r+1}\left[|y-a|^{r+1} + |x-y|^{r+1}\right]\|P_n(\cdot,x)\|_{[a,x],\infty}; \\ \frac{H}{(rp+1)^{\frac{1}{p}}}\left[|y-a|^{rp+1} + |x-y|^{rp+1}\right]^{\frac{1}{p}}\|P_n(\cdot,x)\|_{[a,x],q}, \quad p>1, \\ \qquad\qquad \frac{1}{p}+\frac{1}{q}=1; \\ H\max\{|y-a|^r, |x-y|^r\}\|P_n(\cdot,x)\|_{[a,x],1}. \end{cases} \tag{2.8}$$

Proof. As $f^{(n+1)}$ is of Hölder type, we have

$$\begin{aligned}|S_n(f;a,x,y)| &\leq \left|\int_a^x \left||P_n(t,x)|[f^{(n+1)}(t) - f^{(n+1)}(y)]\right|dt\right| \\ &\leq H\left|\int_a^x |P_n(t,x)||t-y|^r dt\right| \\ &=: A(x,a).\end{aligned}$$

Obviously,

$$\begin{aligned}A(x,a) &\leq H\|P_n(\cdot,x)\|_{[a,x],\infty}\left|\int_a^x |t-y|^r dt\right| \\ &= H\|P_n(\cdot,x)\|_{[a,x],\infty}\frac{|y-a|^{r+1} + |x-y|^{r+1}}{r+1},\end{aligned}$$

which proves the first part of (2.8). Using Hölder's integral inequality, we have:

$$\begin{aligned}A(x,a) &\leq H\|P_n(\cdot,x)\|_{[a,x],q}\left|\int_a^x |t-y|^{rp} dt\right|^{\frac{1}{p}} \\ &= H\|P_n(\cdot,x)\|_{[a,x],q}\left[\frac{|y-a|^{rp+1} + |x-y|^{rp+1}}{rp+1}\right]^{\frac{1}{p}} \\ &= \frac{H}{(rp+1)^{\frac{1}{p}}}\|P_n(\cdot,x)\|_{[a,x],q}[|y-a|^{rp+1} + |x-y|^{rp+1}]^{\frac{1}{p}}\end{aligned}$$

and the second inequality in (2.8) is also proved.

Finally, we have

$$\begin{aligned} A(x,a) &\le H \max_{\substack{t\in[a,x] \\ (t\in[x,a])}} |t-y|^r \cdot \|P_n(\cdot,x)\|_{[a,x],1} \\ &\le H \max\{|y-a|^r, |x-y|^r\} \|P_n(\cdot,x)\|_{[a,x],1} \end{aligned}$$

and the corollary is proved.

Remark 2. We have the following inequalities

$$\begin{aligned} &\left| f(x) - \left[f(a) + \sum_{k=1}^{n} (-1)^{k+1} \left[P_k(x,x) f^{(k)}(x) - P_k(a,x) f^{(k)}(a) \right] \right.\right. \\ &\qquad \left.\left. + (-1)^n [P_{n+1}(x,x) - P_{n+1}(a,x)] f^{(n+1)}(a) \right] \right| \\ &\le B(x,a) := \begin{cases} \frac{H|x-a|^{r+1}}{r+1} \|P_n(\cdot,x)\|_{[a,x],\infty}; \\ \frac{H|x-a|^{r+\frac{1}{p}}}{(rp+1)^{\frac{1}{p}}} \|P_n(\cdot,x)\|_{[a,x],q}, \quad p>1,\ \frac{1}{p}+\frac{1}{q}=1; \\ H|x-a|^r \|P_n(\cdot,x)\|_{[a,x],1}; \end{cases} \end{aligned} \tag{2.9}$$

$$\begin{aligned} &\left| f(x) - \left[f(a) + \sum_{k=1}^{n} (-1)^{k+1} [P_k(x,x) f^{(k)}(x) - P_k(a,x) f^{(k)}(a)] \right.\right. \\ &\qquad \left.\left. + (-1)^n [P_{n+1}(x,x) - P_{n+1}(a,x)] f^{(n+1)}(x) \right] \right| \\ &\le B(x,a); \end{aligned} \tag{2.10}$$

$$\begin{aligned} &\left| f(x) - \left[f(a) + \sum_{k=1}^{n} (-1)^{k+1} [P_k(x,x) f^{(k)}(x) - P_k(a,x) f^{(k)}(a)] \right.\right. \\ &\qquad \left.\left. + (-1)^n [P_{n+1}(x,x) - P_{n+1}(a,x)] f^{(n+1)}\left(\frac{a+x}{2}\right) \right] \right| \\ &\le \frac{1}{2^r} B(x,a), \end{aligned} \tag{2.11}$$

where f and P_n are as in Corollary 3.

3. Some Particular Cases

It is natural now to investigate the case when $f^{(n+1)}$ is assumed to be absolutely continuous.

The following result holds.

Corollary 4. *Let $P_n(t,x)$, f, a, x be as in Theorem 1 and $f^{(n+1)}$ is absolutely continuous on I. Then we have the inequality*

$$|S_n(f;a,x,y)| \le \begin{cases} \left[\frac{1}{4}(x-a)^2 + \left(y-\frac{a+x}{2}\right)^2\right] \|P_n(\cdot,x)\|_{[a,x],\infty}\|f^{(n+2)}\|_{[a,x],\infty}, \\ \quad \text{if } f^{(n+2)} \in L_\infty[a,x]; \\ \left[\frac{|y-a|^{\beta+1}+|x-y|^{\beta+1}}{\beta+1}\right]^{\frac{1}{\beta}} \|P_n(\cdot,x)\|_{[a,x],\alpha}\|f^{(n+2)}\|_{[a,x],\infty}, \\ \quad \text{if } f^{(n+2)} \in L_\infty[a,x]; \quad \alpha>1,\ \frac{1}{\alpha}+\frac{1}{\beta}=1; \\ \left[\frac{1}{2}|x-a| + |y-\frac{a+x}{2}|\right] \|P_n(\cdot,x)\|_{[a,x],1}\|f^{(n+2)}\|_{[a,x],\infty} \\ \quad \text{if } f^{(n+2)} \in L_\infty[a,x]; \\ \left[\frac{|x-y|^{\frac{1}{q}+1}+|y-a|^{\frac{1}{q}+1}}{\frac{1}{q}+1}\right] \|P_n(\cdot,x)\|_{[a,x],\infty}\|f^{(n+2)}\|_{[a,x],p}, \\ \quad p>1,\ \frac{1}{p}+\frac{1}{q}=1, \quad f^{(n+2)} \in L_p[a,x]; \\ \left[\frac{|x-y|^{\frac{\delta}{q}+1}+|y-a|^{\frac{\delta}{q}+1}}{\frac{\delta}{q}+1}\right]^{\frac{1}{\delta}} \|P_n(\cdot,x)\|_{[a,x],\gamma}\|f^{(n+2)}\|_{[a,x],p}, \\ \quad p>1,\ \frac{1}{p}+\frac{1}{q}=1,\ \delta>1,\ \ \frac{1}{\delta}+\frac{1}{\gamma}=1,\ f^{(n+2)} \in L_p[a,x]; \\ \left[\frac{1}{2}|x-a| + |y-\frac{a+x}{2}|\right]^{\frac{1}{\delta}} \|P_n(\cdot,x)\|_{[a,x],1}\|f^{(n+2)}\|_{[a,x],p}, \\ \quad p>1,\ \frac{1}{p}+\frac{1}{q}=1, \quad f^{(n+2)} \in L_p[a,x]; \\ \|P_n(\cdot,x)\|_{[a,x],1}\|f^{(n+2)}\|_{[a,x],1}. \end{cases}$$

Proof. Since $f^{(n+1)}$ is absolutely continuous, then $f^{(n+1)}$ is a.e. differentiable on

$$f^{(n+1)}(t) - f^{(n+1)}(y) = \int_y^t f^{(n+2)}(u)du.$$

We have

$$|f^{(n+1)}(t) - f^{(n+1)}(y)| \le \begin{cases} |t-y|\|f^{(n+2)}\|_{[y,t],\infty}, & \text{if } f^{(n+2)} \in L_\infty[a,x]; \\ |t-y|^{\frac{1}{q}}\|f^{(n+2)}\|_{[y,t],p}, & \text{if } f^{(n+2)} \in L_p[a,x], \\ & \quad p>1,\ \frac{1}{p}+\frac{1}{q}=1; \\ \|f^{(n+2)}\|_{[y,t],1}. \end{cases}$$

Using the representation (2.6), we may get that

$$|S_n(f;a,x,y)| \le \begin{cases} \left|\int_a^x |P_n(t,x)||t-y|\|f^{(n+2)}\|_{[y,t],\infty}dt\right| =: M_1(a,x); \\ \left|\int_a^x |P_n(t,x)||t-y|^{\frac{1}{q}}\|f^{(n+2)}\|_{[y,t],p}dt\right| =: M_2(a,x); \\ \left|\int_a^x |P_n(t,x)|\|f^{(n+2)}\|_{[y,t],1}dt\right| =: M_3(a,x). \end{cases}$$

Since

$$\|f^{(n+2)}\|_{[y,t],\infty} \le \|f^{(n+2)}\|_{[a,x],\infty},$$

we get

$$
\begin{aligned}
&M_1(a,x) \\
&\leq \|f^{(n+2)}\|_{[a,x],\infty}\left|\int_a^x |P_n(t,x)||t-y|dt\right| \\
&\leq \|f^{(n+2)}\|_{[a,x],\infty} \times \begin{cases} \|P_n(\cdot,x)\|_{[a,x],\infty}\left[\frac{1}{4}(x-a)^2+\left(y-\frac{a+x}{2}\right)^2\right]; \\ \|P_n(\cdot,x)\|_{[a,x],\alpha}\left[\frac{|y-a|^{\beta+1}+|x-y|^{\beta+1}}{\beta+1}\right]^{\frac{1}{\beta}}, \\ \qquad \alpha>1,\ \frac{1}{\alpha}+\frac{1}{\beta}=1; \\ \|P_n(\cdot,x)\|_{[a,x],1}\left[\frac{1}{2}|x-a|+|y-\frac{a+x}{2}|\right]. \end{cases}
\end{aligned}
$$

Since

$$\|f^{(n+2)}\|_{[y,t],p} \leq \|f^{(n+2)}\|_{[a,x],p},$$

we get

$$
\begin{aligned}
&M_2(a,x) \\
&\leq \|f^{(n+2)}\|_{[a,x],p}\left|\int_a^x |P_n(t,x)||t-y|^{\frac{1}{q}}dt\right| \\
&\leq \|f^{(n+2)}\|_{[a,x],p} \times \begin{cases} \|P_n(\cdot,x)\|_{[a,x],\infty}\left[\frac{|x-y|^{\frac{1}{q}+1}+|y-a|^{\frac{1}{q}+1}}{\frac{1}{q}+1}\right]; \\ \|P_n(\cdot,x)\|_{[a,x],\gamma}\left[\frac{|x-y|^{\frac{\delta}{q}+1}+|y-a|^{\frac{\delta}{q}+1}}{\frac{\delta}{q}+1}\right]^{\frac{1}{\delta}}, \\ \qquad \gamma>1,\ \frac{1}{\delta}+\frac{1}{\gamma}=1; \\ \|P_n(\cdot,x)\|_{[a,x],1}\left[\frac{1}{2}|x-a|+|y-\frac{a+x}{2}|\right]^{\frac{1}{q}} \end{cases}
\end{aligned}
$$

and, since

$$\|f^{(n+2)}\|_{[y,t],1} \leq \|f^{(n+2)}\|_{[a,x],1},$$

we have

$$M_3(a,x) \leq \|f^{(n+2)}\|_{[a,x],1}\|P_n(\cdot,x)\|_{[a,x],1}$$

and the corollary is proved.

Remark 3. We have the following inequality

$$\begin{aligned}
&\Bigg|f(x)-\Bigg[f(a)+\sum_{k=1}^{n}(-1)^{k+1}[P_k(x,x)f^{(k)}(x)-P_k(a,x)f^{(k)}(a)]\\
&\quad+(-1)^n[P_{n+1}(x,x)-P_{n+1}(a,x)]f^{(n+1)}(a)\Bigg]\Bigg|\\
&\le L(x,a)
\end{aligned}$$

$$:=\begin{cases}
\frac{1}{2}(x-a)^2\|P_n(\cdot,x)\|_{[a,x],\infty}\|f^{(n+2)}\|_{[a,x],\infty}, & \text{if } f^{(n+2)}\in L_\infty[a,x];\\
\frac{1}{(\beta+1)^{\frac{1}{\beta}}}|x-a|^{1+\frac{1}{\beta}} & \\
\quad\times\|P_n(\cdot,x)\|_{[a,x],\alpha}\|f^{(n+2)}\|_{[a,x],\infty}, & \text{if } f^{(n+2)}\in L_\infty[a,x];\\
 & \text{and } \alpha>1,\ \frac{1}{\alpha}+\frac{1}{\beta}=1;\\
|x-a|\|P_n(\cdot,x)\|_{[a,x],1}\|f^{(n+2)}\|_{[a,x],\infty} & \text{if } f^{(n+2)}\in L_\infty[a,x];\\
\frac{q|x-a|^{\frac{1}{q}+1}}{q+1}\|P_n(\cdot,x)\|_{[a,x],\infty}\|f^{(n+2)}\|_{[a,x],p}, & p>1,\ \frac{1}{p}+\frac{1}{q}=1,\\
 & f^{(n+2)}\in L_p[a,x];\\
\frac{q^{\frac{1}{\delta}}|x-a|^{\frac{1}{q}+\frac{1}{\delta}}}{(\delta+q)^{\frac{1}{\delta}}}\|P_n(\cdot,x)\|_{[a,x],\gamma}\|f^{(n+2)}\|_{[a,x],p}, & p>1,\ \frac{1}{p}+\frac{1}{q}=1,\ \delta>1,\\
 & \frac{1}{\delta}+\frac{1}{\gamma}=1,\ f^{(n+2)}\in L_p[a,x];\\
|x-a|^{\frac{1}{q}}\|P_n(\cdot,x)\|_{[a,x],1}\|f^{(n+2)}\|_{[a,x],p}, & p>1,\ \frac{1}{p}+\frac{1}{q}=1,\\
 & f^{(n+2)}\in L_p[a,x];
\end{cases}$$

$$\begin{aligned}
&\Bigg|f(x)-\Bigg[f(a)+\sum_{k=1}^{n}(-1)^{k+1}[P_k(x,x)f^{(k)}(x)-P_k(a,x)f^{(k)}(a)]\\
&\quad+(-1)^n[P_{n+1}(x,x)-P_{n+1}(a,x)]f^{(n+1)}(x)\Bigg]\Bigg|\\
&\le L(x,a);
\end{aligned}$$

$$\begin{aligned}
&\Bigg|f(x)-\Bigg[f(a)+\sum_{k=1}^{n}(-1)^{k+1}[P_k(x,x)f^{(k)}(x)-P_k(a,x)f^{(k)}(a)]\\
&\quad+(-1)^n[P_{n+1}(x,x)-P_{n+1}(a,x)]f^{(n+1)}\Big(\frac{a+x}{2}\Big)\Bigg]\Bigg|\\
&\le\frac{1}{2}L(x,a).
\end{aligned}$$

If we consider the polynomial $P_n(t,x):=\frac{1}{n!}(t-x)^n$, then we have, by (2.5), the following perturbed version of Taylor's formula

$$f(x)=f(a)+\sum_{k=1}^{n}\frac{(x-a)^k}{k!}f^{(k)}(a)+\frac{(x-a)^{n+1}}{(n+1)!}f^{(n+1)}(y)+S_n^T(f;a,x,y), \tag{3.1}$$

where the remainder $S_n^T(f;a,x,y)$ is given by

$$S_n^T(f;a,x,y)=\frac{(-1)^n}{n!}\int_a^x(t-x)^n[f^{(n+1)}(t)-f^{(n+1)}(y)]dt.$$

We have

$$\|P_n(\cdot,x)\|_{[a,x],1} = \frac{1}{n!}\left|\int_a^x |t-x|^n dt\right| = \frac{|x-a|^{n+1}}{(n+1)!},$$

$$\|P_n(\cdot,x)\|_{[a,x],q} = \frac{1}{n!}\left|\int_a^x |t-x|^{nq} dt\right|^{\frac{1}{q}} = \frac{|x-a|^{n+\frac{1}{q}}}{n!(nq+1)^{\frac{1}{q}}}, \quad q\in(1,\infty),$$

$$\|P_n(\cdot,x)\|_{[a,x],\infty} = \frac{1}{n!}\sup_{\substack{t\in[a,x]\\(t\in[x,a])}} |t-x|^n = \frac{|x-a|^n}{n!}, \quad a,x\in I.$$

Using (2.8), we may state that

$$(3.2)\qquad |S_n^T(f;a,x,y)| \le \begin{cases} \frac{H}{(r+1)(n+1)!}|x-a|^{n+1}[|y-a|^{r+1}+|x-y|^{r+1}]; \\ \frac{H|x-a|^{n+\frac{1}{q}}}{n!(rp+1)^{\frac{1}{p}}(nq+1)^{\frac{1}{q}}}[|y-a|^{rp+1}+|x-y|^{rp+1}]^{\frac{1}{p}}, \quad p>1,\ \frac{1}{p}+\frac{1}{q}=1; \\ H\frac{|x-a|^n}{n!}\max\{|y-a|^r,|x-y|^r\}. \end{cases}$$

for all y between x and a.

The following particular inequalities obtained by simply evaluating $A(x,a)$ (defined in Corollary 2) are useful:

$$(3.3)\qquad \left|f(x)-\left[f(a)+\sum_{k=1}^{n+1}\frac{(x-a)^k}{k!}f^{(n)}(a)\right]\right| \le C(x,a) := \begin{cases} \frac{H|x-a|^{n+r+2}}{(r+1)(n+1)!}; \\ \frac{H|x-a|^{n+r+1}}{n!(rp+1)^{\frac{1}{p}}(nq+1)^{\frac{1}{q}}}; \\ H\frac{|x-a|^{n+r}}{n!} \end{cases}$$

provided that $f^{(n+1)}$ is of r-H-Hölder type on I.

Example 1. We also have

$$(3.4)\qquad \left|f(x)-\left[f(a)+\sum_{k=1}^{n}\frac{(x-a)^k}{k!}f^{(k)}(a)+\frac{(x-a)^{n+1}}{(n+1)!}f^{(n+1)}(x)\right]\right| \le C(x,a)$$

and

$$(3.5)\qquad \left|f(x)-\left[f(a)+\sum_{k=1}^{n}\frac{(x-a)^k}{k!}f^{(k)}(a)+\frac{(x-a)^{n+1}}{(n+1)!}f^{(n+1)}\left(\frac{a+x}{2}\right)\right]\right| \le \frac{1}{2^r}C(x,a),$$

where $f^{(n+1)}$ is of $r-H-$Hölder type on I.

If we consider the polynomial $P_n(t,x) := \frac{1}{n!}\left(t-\frac{a+x}{2}\right)^n$, then, by (2.5), we may state:

$$f(x) = f(a)+\sum_{k=1}^{n}\frac{(x-a)^k}{2^k k!}\left[f^{(k)}(a)+(-1)^{k+1}f^{(k)}(x)\right] + \frac{[1+(-1)^n](x-a)^{n+1}}{2^{n+1}(n+1)}f^{(n+1)}(y)+S_n^M(f;a,x,y),$$

where the remainder $S_n^M(f;a,x,y)$ is given by

$$S_n^M(f;a,x,y) = \frac{(-1)^n}{n!}\int_a^x (t-\frac{a+x}{2})^n[f^{(n+1)}(t)-f^{(n+1)}(y)]dt.$$

We have

$$\begin{aligned}
\|P_n(\cdot,x)\|_{[a,x],1} &= \frac{1}{n!}\left|\int_a^x \left|t-\frac{a+x}{2}\right|^n dt\right| \\
&= \frac{1}{n!}\left|\int_a^{\frac{a+x}{2}} \left|t-\frac{a+x}{2}\right|^n dt + \int_{\frac{a+x}{2}}^x \left|t-\frac{a+x}{2}\right|^n dt\right| \\
&= \frac{1}{n!(n+1)}\left|\frac{|x-a|^{n+1}}{2^{n+1}}+\frac{|x-a|^{n+1}}{2^{n+1}}\right| \\
&= \frac{|x-a|^{n+1}}{(n+1)!2^n};
\end{aligned}$$

$$\begin{aligned}
\|P_n(\cdot,x)\|_{[a,x],q} &= \frac{1}{n!}\left|\int_a^x \left|t-\frac{a+x}{2}\right|^{nq} dt\right|^{\frac{1}{q}} \\
&= \frac{1}{n!}\left|\int_a^{\frac{a+x}{2}} \left|t-\frac{a+x}{2}\right|^{nq} dt + \int_{\frac{a+x}{2}}^x \left|t-\frac{a+x}{2}\right|^{nq} dt\right|^{\frac{1}{q}} \\
&= \frac{1}{n!}\left[\frac{|x-a|^{nq+1}}{(nq+1)2^{nq+1}}+\frac{|x-a|^{nq+1}}{(nq+1)2^{nq+1}}\right]^{\frac{1}{q}} \\
&= \frac{1}{2^n n!(nq+1)^{\frac{1}{q}}}|x-a|^{n+\frac{1}{q}};
\end{aligned}$$

$$\|P_n(\cdot,x)\|_{[a,x],\infty} = \frac{1}{n!}\sup_{t\in[a,x]}\left|t-\frac{a+x}{2}\right|^n = \frac{|x-a|^n}{2^n n!}.$$

Using (2.8), we may state that

$$|S_n^M(f;a,x,y)| \le \begin{cases} \frac{H|x-a|^{n+1}}{(n+1)!2^n(r+1)}[|y-a|^{r+1}+|x-y|^{r+1}]; \\ \frac{H|x-a|^{n+\frac{1}{q}}}{(rp+1)^{\frac{1}{p}}(nq+1)^{\frac{1}{q}}2^n n!}[|y-a|^{rp+1}+|x-y|^{rp+1}]^{\frac{1}{p}}, \\ \qquad p>1,\ \frac{1}{p}+\frac{1}{q}=1; \\ H\frac{|x-a|^n}{2^n n!}\max\{|y-a|^r,|x-y|^r\}. \end{cases} \tag{3.7}$$

The following particular inequalities which follow by directly evaluating $A(x,a)$ are useful

$$\begin{aligned}
&\left|f(x)-\left[f(a)+\sum_{k=1}^n \frac{(x-a)^k}{2^k k!}[f^{(k)}(a)+(-1)^{k+1}f^{(k)}(x)]\right.\right. \\
&\left.\left.+\frac{[1+(-1)^n](x-a)^{n+1}}{2^{n+1}(n+1)}f^{(n+1)}(a)\right]\right| \\
&\le D(x,a) := \begin{cases} \frac{H|x-a|^{n+r+2}}{(n+1)!2^n(r+1)}; \\ \frac{H|x-a|^{n+r+1}}{(rp+1)^{\frac{1}{p}}(nq+1)^{\frac{1}{q}}2^n n!}, \quad p>1,\ \frac{1}{p}+\frac{1}{q}=1; \\ H\frac{|x-a|^{n+r}}{2^n n!}, \end{cases}
\end{aligned} \tag{3.8}$$

where $f^{(n+1)}$ is of $r-H-$Hölder type on I.

We also have

$$\begin{aligned}(3.9)\qquad &\Bigg|f(x)-\Bigg[f(a)+\sum_{k=1}^{n}\frac{(x-a)^k}{2^k k!}[f^{(k)}(a)+(-1)^{k+1}f^{(k)}(x)]\\&+\frac{[1+(-1)^n](x-a)^{n+1}}{2^{n+1}(n+1)}f^{(n+1)}(x)\Bigg]\Bigg|\\&\le D(x,a);\end{aligned}$$

$$\begin{aligned}(3.10)\qquad &\Bigg|f(x)-\Bigg[f(a)+\sum_{k=1}^{n}\frac{(x-a)^k}{2^k k!}[f^{(k)}(a)+(-1)^{k+1}f^{(k)}(x)]\\&+\frac{[1+(-1)^n](x-a)^{n+1}}{2^{n+1}(n+1)}f^{(n+1)}(\frac{a+x}{2})\Bigg]\Bigg|\\&\le\frac{1}{2^r}D(x,a),\quad a,x\in I.\end{aligned}$$

Remark 4. Similar inequalities may be stated if we use Corollary 4.

Remark 5. Similar inequalities can be stated for the other choices of "harmonic polynomials" $P_n(t,x)$ considered in [4]. As we do not have explicit or at least bounds for $\|P_n(\cdot,x)\|_p$, $p\in[1,\infty]$, we omit the details.

The following corollaries are given for particular Peano kernels based on the order $a<y<x$.

Under the assumptions of Theorem 2, let a point $y\in I\subset\mathbb{R}$ such that $a<y<x$. If we consider the polynomial $P_n(t,x):=\frac{1}{n!}(t-x)^n$, then, by (2.5), we have the perturbed version of Taylor's formula

$$(3.11)\qquad f(x)=f(a)+\sum_{k=1}^{n}\frac{(x-a)^k}{k!}f^{(k)}(a)+\frac{(x-a)^{n+1}}{(n+1)!}f^{(n+1)}(y)+\tilde{S}_n^T(f;a,y,x),$$

where the remainder $\tilde{S}_n^T(f;a,y,x)$ is given by

$$(3.12)\qquad \tilde{S}_n^T(f;a,y,x):=\frac{(-1)^n}{n!}\int_a^x(t-x)^n[f^{(n+1)}(t)-f^{(n+1)}(y)]dt.$$

The following corollary holds for functions f for which $f^{(n+1)}$ is Hölder continuous.

Corollary 5. *If the function $f^{(n+1)}:I\to\mathbb{R}$ is of r-H-Hölder type on I, we have, for $H>0$ and $r\in[0,1]$, the bound*

$$(3.13)\qquad |\tilde{S}_n^T(f;a,y,x)|\le\frac{H}{n!}(x-y)^{n+r+1}[\Psi_\alpha(n+1,r+1)+B(n+1,r+1)],$$

where

$$\alpha=\frac{y-a}{x-y}>0,\quad B(l,s)=\int_0^1 t^{l-1}(1-t)^{s-1}dt,\quad l,s>0,$$

is the classical Beta function and

$$\Psi_r(l,s)=\int_0^r t^{l-1}(1+t)^{s-1}dt,\quad l,s>0,$$

is a real positive valued integral.

Proof. As $f^{(n+1)}$ is of Hölder type, we have

$$|\tilde{S}_n^T(f;a,y,x)| \leq \frac{H}{n!}\int_a^x |t-x|^n |t-y|^r dt = \tilde{A}(a,x),$$

where

$$\tilde{A}(a,x) = \frac{H}{n!}\left[\int_a^y (x-t)^n (y-t)^r dt + \int_y^x (x-t)^n (t-y)^r dt\right].$$

Making the substitutions

$$t = y-(x-y)w \quad \text{and} \quad t = y+(x-y)w$$

into the first and second integrals respectively, we obtain

$$A(a,x) = \frac{H}{n!}(x-y)^{n+r+1}\left[\Psi_\alpha(n+1,r+1) + B(n+1,r+1)\right]$$

and so Corollary 5 follows.

Remark 6. When $y=a$ the following inequality holds provided that $f^{(n+1)}$ is of r-H-Hölder type on I

$$\left|f(x) - \sum_{k=0}^{n+1}\frac{(x-a)^k}{k!}f^{(n)}(a)\right| \leq \frac{H(x-a)^{n+r+1}}{n!}B(n+1,r+1).$$

Corollary 6. *If $f^{(n+1)}: I \to \mathbb{R}$ is absolutely continuous on I, then we have*

$$\left|f(x) - f(a) - \sum_{k=1}^{n}\frac{(x-a)^k}{k!}f^{(k)}(a) - \frac{(x-a)^{n+1}}{(n+1)!}f^{(n+1)}(y)\right|$$
$$< \|f^{(n+2)}\|_\infty\left(\frac{(y-a)(x-a)^{n+1}}{(n+1)!} - \frac{(x-a)^{n+2}}{(n+2)!} + \frac{2(x-y)^{n+2}}{(n+2)!}\right),$$

where $f^{n+2} \in L_\infty[a,x]$.

Proof. From Corollary 4, we have

$$\begin{aligned}\int_a^x |P_n(t,x)||t-y|dt &= \int_a^y \frac{|t-x|^n|t-y|}{n!}dt + \int_y^x \frac{|t-x|^n|t-y|}{n!}dt\\ &= \int_a^y \frac{(x-t)^n(y-t)}{n!}dt + \int_y^x \frac{(x-t)^n(t-y)}{n!}dt\\ &= \frac{(y-a)(x-a)^{n+1}}{(n+1)!} - \frac{(x-a)^{n+2}}{(n+2)!} + \frac{2(x-y)^{n+2}}{(n+2)!}\end{aligned}$$

and so Corollary 6 follows.

Now, consider the polynomial

$$P_n(t,x) := \frac{1}{n!}(t-\frac{a+x}{2})^n$$

and, from (2.5), we may state

$$\begin{aligned}
M &:= f(x) - f(a) - \sum_{k=1}^{n} \frac{(x-a)^k}{2^k k!}[f^{(k)}(a) + (-1)^{k+1} f^{(k)}(x)] \\
&\quad - \frac{[1+(-1)^n](x-a)^{n+1}}{2^{n+1}(n+1)!} f^{(n+1)}(y) \\
&= \tilde{S}_n^M(f; a, y, x),
\end{aligned}$$

where the remainder $\tilde{S}_n^M(f; a, y, x)$ is given by

$$\tilde{S}_n^M(f; a, y, x) = \frac{(-1)^n}{n!} \int_a^x \left(t - \frac{a+x}{2}\right)^n [f^{(n+1)}(t) - f^{(n+1)}(y)]dt.$$

Corollary 7. *If $f^{(n+1)} : I \to \mathbb{R}$ is absolutely continuous on I and $a < y < x$, then we have*

$$|\tilde{S}_n^M(f; a, y, x)| \leq \frac{2\|f^{(n+2)}\|_\infty}{(n+1)!} \left(\frac{x-a}{2}\right)^{n+1} \left|y - \frac{a+x}{2}\right|.$$

Proof. Consider

$$I = \int_a^x \frac{1}{n!}\left|t - \frac{a+x}{2}\right|^n |t-y| dt.$$

If $\frac{a+x}{2} \in (a, y)$, then

$$\begin{aligned}
n! I_1 &= \int_a^{\frac{a+x}{2}} \left(\frac{a+x}{2} - t\right)^n (y-t)dt + \int_{\frac{a+x}{2}}^{y} \left(t - \frac{a+x}{2}\right)^n (y-t)dt \\
&\quad + \int_y^x \left(t - \frac{a+x}{2}\right)^n (y-t)dt \\
&= \int_a^{\frac{a+x}{2}} \left(\frac{a+x}{2} - t\right)^n (y-t)dt + \int_{\frac{a+x}{2}}^{x} \left(t - \frac{a+x}{2}\right)^n (y-t)dt.
\end{aligned}$$

Integration by parts leads to the result

$$I_1 = \frac{2}{(n+1)!}\left(\frac{x-a}{2}\right)^{n+1}\left(y - \frac{a+x}{2}\right).$$

If $\frac{a+x}{2} \in [y, x)$, then also

$$\begin{aligned}
n! I_2 &= \int_a^{y} \left(\frac{a+x}{2} - t\right)^n (t-y)dt + \int_{y}^{\frac{a+x}{2}} \left(\frac{a+x}{2} - t\right)^n (t-y)dt \\
&\quad + \int_{\frac{a+x}{2}}^x \left(t - \frac{a+x}{2}\right)^n (t-y)dt \\
&= \int_a^{\frac{a+x}{2}} \left(\frac{a+x}{2} - t\right)^n (t-y)dt + \int_{\frac{a+x}{2}}^{x} \left(t - \frac{a+x}{2}\right)^n (t-y)dt.
\end{aligned}$$

Integration by parts leads to the result

$$I_2 = \frac{2}{(n+1)!}\left(\frac{x-a}{2}\right)^{n+1}\left(-y + \frac{a+x}{2}\right).$$

Utilising I_1 and I_2, we see that Corollary 7 follows.

Remark 7. When $y = a$, the following inequality holds

$$\Big| f(x) - f(a) - \sum_{k=1}^{n} \frac{(x-a)^k}{2^k k!}[f^{(k)}(a) + (-1)^{k+1} f^{(k)}(x)] - \frac{[1+(-1)^n](x-a)^{n+1}}{2^{n+1}(n+1)!} f^{(n+1)}(a) \Big| < \frac{2\|f^{(n+2)}\|_\infty}{(n+1)!}\Big(\frac{x-a}{2}\Big)^{n+2}.$$

References

1. G. A. Anastassiou and S. S. Dragomir, *On some estimates of the remainder in Taylor's formula*, J. of Math. Anal. Appl. **263** (2001), 246–263.
2. P. Appell, *Sur une classe de polnômes*, Ann. Sci. Ecole Norm. Sup. **9(2)** (1880), 119–144.
3. S. S. Dragomir, *An improvement of the remainder estimate in the generalized Taylor formula*, RGMIA Res. Rep. Coll. **3(1)** (2000); *Article 1.*
4. M. Matić, J. E. Pečarić and N. Ujević, *On new estimation of the remainder in generalized Taylor's formula*, Math. Ineq. Appl. **2(3)** (1999), 343–361.

In: Inequality Theory and Applications, Volume 4
Editors: Yeol Je Cho et al., pp.85–92

ISBN 1-59454-874-9

CHARACTERISTIC PROPERTY FOR INEQUALITIES OF BOUNDED LINEAR OPERATORS

C.-S. LIN[1] AND Y. J. CHO[2]

ABSTRACT. In recent years it has been proved by many authors that many well known and useful operator inequalities on a Hilbert space in the past are equivalent to the Löwner-Heinz formula, which is a fundamental one in theory of operator inequalities. In this article we prove that such well known operator inequalities could be generalized in such a way that they are all equivalent to the Furuta inequality. After all, the Furuta inequality is a remarkable generalization of the Löwner-Heinz formula. Finally, we give some new characterizations of the Heinz-Kato inequality.

1. Introduction

Throughout this note it is to be understood that the capital letters mean bounded linear operators acting on a Hilbert space H. T is positive (written $T \geq O$, the zero operator) in case $(Tx, x) \geq 0$ for all $x \in$H, and the notation $T > O$ means T is positive and invertible. If S and T are selfadjoint, we write $T \geq S$ in case $T - S \geq O$. I denotes the identity operator. $T = U|T|$ is the polar decomposition of T with U the partial isometry, where $|T|$ is the positive square root of the positive operator T^*T, whereas T^* denotes the adjoint of T.

In recent years, many authors have been interested in the classical Löwner-Heinz formula [9, 15], a monotone function of positive operators which is undoubtedly fundamental in theory of operator inequality on H. To begin with, a remarkable generalization of the Löwner-Heinz formula was discovered and given by Furuta in 1987 [4]. Subsequently, a lot of papers have been written and proved that the Löwner-Heinz formula is indeed equivalent to many well known and useful operator inequalities and norm inequalities of operators in the past, such as the Cordes inequality [1], the McIntosh inequality [16], Reid's inequality [17], and many more other inequalities [2, 3, 5, 6, 11, 12, 13, 14]. Our work in this article will be concerned primarily with generalizing these operator inequalities in such a way that they are all euivalent to the Furuta inequality. This is indeed a natural approach from the above point of view. In the last section we give some new characterizations of the Heinz-Kato inequality. Notice that normally such characterizations are involved with polar decompositions of operators in the literature [6, 12, 14 and references cited in there], but

---

Received November 24, 2003.
2000*Mathematics Subject Classification*. Primary 47A63, 47A30.
Key words and phrases. Löwner-Heinz formula, Löwner inequality, contraction operator, Furuta inequality, Cordes inequality, Jensen-Hansen inequality, McIntosh inequality, Reid's inequality, Heinz-Kato inequality.

[1] Department of Mathematics, Bishop's University, Lennoxville, Quebec, J1M 1Z7, Canada. E-mail: plin@ubishops.ca

[2] Department of Mathematics, Gyeongsang National university, Chinju 660-701, Korea. E-mail: yjcho@nongae.gsnu.ac.kr

ours are different. It should be mentioned that recent contributions in the literature to the Furuta inequality and its generalization have brought substantial applications in general theory of operator inequalities on H. And the most recent book in [7] is excellent resources for the Furuta inequality, operator inequalities in general, and its related topics.

First of all, let us recall the Löwner-Heinz formula [9]: The operator inequality $A^\alpha \geq B^\alpha$ holds if $A \geq B \geq O$ and $\alpha \in [0, 1]$. But the inequality does not hold in general for $\alpha > 1$ [7, §3.2, Corollary 1]. Notice that a classical characterization of the Löwner-Heinz formula is as follows: The operator inequality $A^{1/2} \geq B^{1/2}$ holds if $A \geq B \geq O$ [2], which is known as the Löwner inequality [15].

2. Characteristic Property of the Löwner-Heinz Formula

In this section we shall list systematically the most well known recent characterizations of the classical Löwner-Heinz formula obtained from [2, 3, 5, 6, 11, 12, 13, 14]. Thus, the proof will not be given but precise references will be mentioned for the convenience of the reader.

Proposition. *Let $A, B, C, D \geq O, T$ and K be arbitrary operators with AK selfadjoint, and let E be any contraction operator. Then the following are equivalent for $\alpha, \beta \in (0, 1]$, $t \in [1, 2]$, and $\delta \geq 1$.*

(1) $A \geq B \geq O$ *implies* $A^\alpha \geq B^\alpha$ *(Löwner-Heinz formula [9]);*

(1′) $A \geq B \geq O$ *implies* $A^{1/2} \geq B^{1/2}$ *(Löwner inequality [15]);*

(2) $\|A^\alpha B^\alpha\| \leq \|AB\|^\alpha$ *(Cordes inequality [1]);*

(3) $\|A^{1/2}B^{1/2}\| \leq \|AB\|^{1/2}$;

(4) $\|AB\|^\delta \leq \|A^\delta B^\delta\|$;

(5) $\|AB\|^t \leq \|A^t B^t\|$;

(6) $\|AB\|^2 \leq \|A^2B^2\|$;

(7) $\|A^\alpha B^\alpha A^\alpha\| \leq \|ABA\|^\alpha$;

(8) $\|ABA\|^t \leq \|A^t B^t A^t\|$;

(9) $E^*A^\alpha E \leq (E^*AE)^\alpha$ *(Jensen-Hansen inequality [8]);*

(10) $(E^*AE)^t \leq E^*A^tE$;

(11) $\|A^\alpha TB^{1-\alpha}\| \leq \|AT\|^\alpha \|TB\|^{1-\alpha}$ *(McIntosh inequality [16]);*

(11′) $\|ABA\| \leq \|A^2B\|$;

(12) $|(Tx, y)| \leq \|A^\alpha x\|\|B^{1-\alpha}y\|$ *for all* $x, y \in H$ *such that* $T^*T \leq A^2$ *and* $TT^* \leq B^2$ *(Heinz-Kato inequality [10]);*

(13) $|(T|T|^{\alpha+\beta-1}x, y)| \leq \|A^\alpha x\|\|B^\beta y\|$ *for all* $x, y \in H$ *such that* $\alpha + \beta \geq 1$, $T^*T \leq A^2$ *and* $TT^* \leq B^2$ *(extended Heinz-Kato inequality [7]);*

(14) $A \geq B \geq O$ *and* $C \geq D \geq O$ *imply* $\|B^{\alpha/2}TD^{\beta/2}\| \leq \|A^{\alpha/2}TC^{\beta/2}\|$;

(15) $|(AK|AK|^{\alpha+\beta-1}x, y)| \leq \|K\|^{\alpha+\beta}\|A^\alpha x\|\|A^\beta y\|$ *for all* $x, y \in H$ *such that* $\alpha + \beta \geq 1$, *and* $AK = V \mid AK|$ *is the polar decomposition of* AK *with* V *the partial isometry;*

(15′) $|(AKx, x)| \leq \|K\|(Ax, x)$ *for all* $x \in H$ *(Reid's inequality [17]).*

The proof of Proposition can be found in recent articles as follows, and we shall mention some related references only. Inequalities (1) through (10), except (1′) and (4), are equivalent due to [3, Theorem 1]. (1′) can be found in [2, Lemma 1] and (4) is in [5, Theorem]. As for the remainder of inequalities, equivalence of (1), (11), (11′) and (12) were proved in [2, Theorem 6]; (13) is in [14, Corollary 1]; and (14) is a special case of (14) in Theorem 2 in the next section (let $r = s = 0$ in Theorem 2), and we will prove it there. Finally, (15) and

(15′) are due to [13, Corollary 1 and 2]. In fact, two inequalities (15′) and (1′) are equivalent as was proved in [13, Corollary 2].

3. Characteristic Property of the Furuta Inequality

Recall the Furuta inequality from [4]: If $A \geq B \geq O$, then $(B^r A^p B^r)^{\frac{1}{q}} \geq B^{\frac{p+2r}{q}}$ for $p, r \geq 0$, and $q \geq 1$ such that $(1+2r)q \geq p+2r$. Clearly, by letting $r = 0$ we obtain the Löwner-Heinz formula. Notice that the Furuta inequality always holds for $1 \geq p \geq 0$ due to the Löwner-Heinz formula. Instead of this original Furuta inequality we need an altered equivalent form, which is shown in (3) of Theorem 1 below. We require this simple and crucial result in our investigations, which does not seem to have a convenient reference in the literature.

Theorem 1. *Let $A, B \geq O$, also let $r \geq 0$ and $p \geq 1$. Then the following statements hold and are mutually equivalent.*

(1) *$A \geq B \geq O$ implies $(B^r A^p B^r)^{\frac{1}{q}} \geq B^{\frac{p+2r}{q}}$ for $q \geq 1$ such that $(1+2r)q \geq p+2r$ (Furuta inequality);*

(2) *$A \geq B \geq O$ implies $(B^r A^p B^r)^{\frac{(1+2r)\alpha}{p+2r}} \geq B^{(1+2r)\alpha}$ for $\alpha \in (0,1]$;*

(3) *$(B^r A^p B^r)^{\frac{1+2r}{p+2r}} \geq B^{1+2r}$ implies $(B^r A^p B^r)^{\frac{(1+2r)\alpha}{p+2r}} \geq B^{(1+2r)\alpha}$ for $\alpha \in (0,1]$.*

Proof. (1)⇒(2). Let $\alpha = \frac{p+2r}{q(1+2r)} \leq 1$ in (1). Then $\alpha \in (0,1]$ and (2) holds.

(2)⇒(3). The statement (2) means in particular that $A \geq B \geq O$ implies $(B^r A^p B^r)^{\frac{1+2r}{p+2r}} \geq B^{1+2r} \geq O$. Let $r = 0$ in the latter inequality. Then

$$A \geq B \geq O.$$

In other words, under assertion (2), we have $A \geq B \geq O$ if and only if

$$(B^r A^p B^r)^{\frac{1+2r}{p+2r}} \geq B^{1+2r}.$$

Hence (3) follows.

(3)⇒(1). In (3) let $r = 0$. Then (3) means in particular that $A \geq B \geq O$ implies $A^\alpha \geq B^\alpha$ for $\alpha \in (0,1]$. This is precisely the Löwner-Heinz formula in (1) of Proposition, which is required for proving the original Furuta's inequality (see a remark in [7, §3.2, p.150]). Hence we have (1). This completes the proof.

Notice that (3) in Theorem 1 holds true, of course, due to the Löwner-Heinz formula. But we don't assume the formula in the proof. The next result will be used in the proof of Theorem 2 below.

Lemma. [7, Theorem 4, §2.2.2] *Let $T = U|T|$ be the polar decomposition of T. Then, for $\gamma > 0$, we have the equality*

$$|T^*|^\gamma = U|T|^\gamma U^*.$$

We now proceed to derive the main result, i.e., generalizing every inequality in Proposition in such a way that they are all equivalent to one another, and also to the Furuta inequality. Note that some of them are already known, for which we will mention only the related references without proofs.

Theorem 2. *Let $A, B, C, D \geq O$, T and K be arbitrary operators with AK selfadjoint, and let E be any contraction operator. Then the following are equivalent for $\alpha, \beta \in (0,1]$, $r, s \geq 0$, $p, q \geq 1$, $t \in [1,2]$, and $\delta \geq 1$.*

(1) *$A \geq B \geq O$ implies $(B^r A^p B^r)^{\frac{1}{q}} \geq B^{\frac{p+2r}{q}}$ for $(1+2r)q \geq p+2r$ (Furuta inequality);*

(1′) $A \geq B \geq O$ *implies* $(B^rA^pB^r)^{\frac{1+2r}{2(p+2r)}} \geq B^{\frac{1+2r}{2}}$;

(2) $\|(B^rA^pB^r)^{\frac{(1+2r)\alpha}{p+2r}}B^{(1+2r)\alpha}\| \leq \|(B^rA^pB^r)^{\frac{1+2r}{p+2r}}B^{1+2r}\|$;

(3) $\|(B^rA^pB^r)^{\frac{1+2r}{2(p+2r)}}B^{\frac{1+2r}{2}}\| \leq \|(B^rA^pB^r)^{\frac{1+2r}{p+2r}}B^{1+2r}\|^{1/2}$;

(4) $\|(B^rA^pB^r)^{\frac{1+2r}{p+2r}}B^{1+2r}\|^{\delta} \leq \|(B^rA^pB^r)^{\frac{(1+2r)\delta}{p+2r}}B^{(1+2r)\delta}\|$;

(5) $\|(B^rA^pB^r)^{\frac{1+2r}{p+2r}}B^{1+2r}\|^{t} \leq \|(B^rA^pB^r)^{\frac{(1+2r)t}{p+2r}}B^{(1+2r)t}\|$;

(6) $\|(B^rA^pB^r)^{\frac{1+2r}{p+2r}}B^{1+2r}\|^{2} \leq \|(B^rA^pB^r)^{\frac{2(1+2r)}{p+2r}}B^{2(1+2r)}\|$;

(7) $\|(B^rA^pB^r)^{\frac{(1+2r)\alpha}{p+2r}}B^{(1+2r)\alpha}(B^rA^pB^r)^{\frac{(1+2r)\alpha}{p+2r}}\|$
$\leq \|(B^rA^pB^r)^{\frac{1+2r}{p+2r}}B^{1+2r}(B^rA^pB^r)^{\frac{1+2r}{p+2r}}\|^{\alpha}$;

(8) $\|(B^rA^pB^r)^{\frac{1+2r}{p+2r}}B^{1+2r}(B^rA^pB^r)^{\frac{1+2r}{p+2r}}\|^{t}$
$\leq \|(B^rA^pB^r)^{\frac{(1+2r)t}{p+2r}}B^{(1+2r)t}(B^rA^pB^r)^{\frac{(1+2r)t}{p+2r}}\|$;

(9) $E^*(B^rA^pB^r)^{\frac{(1+2r)\alpha}{p+2r}}E \leq [E^*(B^rA^pB^r)^{\frac{1+2r}{p+2r}}E]^{\alpha}$;

(10) $[E^*(B^rA^pB^r)^{\frac{1+2r}{p+2r}}E]^{t} \leq E^*(B^rA^pB^r)^{\frac{(1+2r)t}{p+2r}}E$;

(11) $\|(B^rA^pB^r)^{\frac{(1+2r)\alpha}{p+2r}}TB^{(1+2r)(1-\alpha)}\| \leq \|(B^rA^pB^r)^{\frac{1+2r}{p+2r}}T\|^{\alpha}\|TB^{(1+2r)}\|^{1-\alpha}$;

(11′) $\|(B^rA^pB^r)^{\frac{1+2r}{p+2r}}B^{1+2r}(B^rA^pB^r)^{\frac{1+2r}{p+2r}}\| \leq \|(B^rA^pB^r)^{\frac{2(1+2r)}{p+2r}}B^{1+2r}\|$;

(12) $|(T|T|^{(1+2r)\alpha+(1+2s)\beta-1}x,y)|^2$
$\leq ((|T|^{2r}A^{2p}|T|^{2r})^{\frac{(1+2r)\alpha}{p+2r}}x,x)((|T^*|^{2s}B^{2q}|T^*|^{2s})^{\frac{(1+2s)\beta}{q+2s}}y,y)$ *for all* $x,y \in H$ *such that* $T^*T \leq A^2$ and $TT^* \leq B^2$, and $(1+2r)\alpha+(1+2s)\beta \geq 1$;

(13) $|(|T^*|^{(1+2s)\beta}U|T|^{(1+2r)\alpha}x,y)|$
$\leq \|(|T|^{2r}A^{2p}|T|^{2r})^{\frac{(1+2r)\alpha}{2(p+2r)}}x\|\|(|T^*|^{2s}B^{2q}|T^*|^{2s})^{\frac{(1+2s)\beta}{2(q+2s)}}y\|$ *for all* $x,y \in H$ *such that* $T^*T \leq A^2$ and $TT^* \leq B^2$, and $(1+2r)\alpha+(1+2s)\beta \geq 1$;

(14) $A \geq B \geq O$ and $C \geq D \geq O$ *imply*
$\|B^{(1+2r)\alpha/2}TD^{(1+2s)\beta/2}\| \leq \|(B^rA^pB^r)^{\frac{(1+2r)\alpha}{2(p+2r)}}T(D^sC^qD^s)^{\frac{(1+2s)\beta}{2(q+2s)}}\|$;

(15) $|(AK|AK|^{(1+2r)\alpha+(1+2s)\beta-1}x,y)|^2$
$\leq \|K\|^c((|AK|^{2r}A^{2p}|AK|^{2r})^{\frac{(1+2r)\alpha}{p+2r}}x,x)((|AK|^{2s}A^{2q}|AK|^{2s})^{\frac{(1+2s)\beta}{q+2s}}y,y)$
for all $x,y \in H$, *where* $c = \frac{2p(1+2r)\alpha}{p+2r} + \frac{2q(1+2s)\beta}{q+2s}$ *such that* $(1+2r)\alpha+(1+2s)\beta \geq 1$. *and* $AK = V|AK|$ *is the polar decomposition of* AK *with* V *the partial isometry.*

Proof. We begin with replacing A and B in (1) through (11′) of Proposition by $(B^rA^pB^r)^{\frac{1+2r}{p+2r}}$ and B^{1+2r}, respectively. Thus, immediately we obtain inequalities from (2) to (11′), and they are all equivalent to one another. As for details, due to such replacement (1) and (1′) in Proposition become

$$(B^rA^pB^r)^{\frac{1+2r}{p+2r}} \geq B^{1+2r} \text{ implies } (B^rA^pB^r)^{\frac{(1+2r)\alpha}{p+2r}} \geq B^{(1+2r)\alpha}$$

and

$$(B^rA^pB^r)^{\frac{1+2r}{p+2r}} \geq B^{1+2r} \text{ implies } (B^rA^pB^r)^{\frac{1+2r}{2(p+2r)}} \geq B^{\frac{1+2r}{2}},$$

respectively. With the aid of Theorem 1 it is easily seen that the above two inequalities are all equivalent to the Furuta inequality. Thus, we obtain (1) and (1′). Remark that (1′) is precisely (2) in Theorem 1 when $\alpha = \frac{1}{2}$.

Next, due to Lemma, equivalence of (1), (12), (13) and (15) can be found in [6, Theorem 1], [12, Theorem 2] and [13, Theorem 1], respectively, and the proof should be omitted. It remains to prove equivalence of two statements (1′) (or equivalently, (2) of Theorem 1) and (14) as follows.

(2) of Theorem 1 implies (14). In (2) of Theorem 1 let A be replaced by C, B by D, and let $r = s$, $p = q$ and $\alpha = \beta$. Then

$$(*)\quad C \geq D \geq O \text{ implies } (D^s C^q D^s)^{\frac{(1+2s)\beta}{q+2s}} \geq D^{(1+2s)\beta}.$$

Now,

$$\begin{aligned}
&\|B^{(1+2r)\alpha/2} T D^{(1+2s)\beta/2}\|^2 \\
&= \|B^{(1+2r)\alpha/2} T D^{(1+2s)\beta} T^* B^{(1+2r)\alpha/2}\| \\
&\leq \|B^{(1+2r)\alpha/2} T (D^s C^q D^s)^{\frac{(1+2t)\beta}{q+2t}} T^* B^{(1+2r)\alpha/2}\| \qquad \text{by } (*) \\
&= \|(D^s C^q D^s)^{\frac{(1+2s)\beta}{2(q+2s)}} T^* B^{(1+2r)\alpha/2}\|^2 \\
&= \|(D^s C^q D^s)^{\frac{(1+2s)\beta}{2(q+2s)}} T^* B^{(1+2r)\alpha} T (D^s C^q D^s)^{\frac{(1+2s)\beta}{2(q+2s)}}\| \\
&\leq \|(D^s C^q D^s)^{\frac{(1+2s)\beta}{2(q+2s)}} T^* (B^r A^p B^r)^{\frac{(1+2r)\alpha}{p+2r}} T (D^s C^q D^s)^{\frac{(1+2s)\beta}{2(q+2s)}}\| \\
&\qquad\qquad\qquad\qquad \text{by (2) in Theorem 1} \\
&= \|(B^r A^p B^r)^{\frac{(1+2r)\alpha}{2(p+2r)}} T (D^s C^q D^s)^{\frac{(1+2s)\beta}{2(q+2s)}}\|^2,
\end{aligned}$$

and (14) follows.

(14) implies (2) of Theorem 1. We may assume without loss of generality that $A, B > O$ and let, in particular, $C = D = I$ and $T = (B^r A^p B^r)^{-\frac{(1+2r)\alpha}{2(p+2r)}}$ in (14). Then

$$\|B^{(1+2r)\alpha/2}(B^r A^p B^r)^{-\frac{(1+2r)\alpha}{2(p+2r)}}\| \leq 1,$$

which is the same as the inequality

$$(B^r A^p B^r)^{-\frac{(1+2r)\alpha}{2(p+2r)}} B^{(1+2r)\alpha} (B^r A^p B^r)^{-\frac{(1+2r)\alpha}{2(p+2r)}} \leq I,$$

and hence (2) of Theorem 1 follows. This completes the proof.

Remark that (1′), (2), (9), (11), (12) and (15) in Theorem 2 are generalizations of inequalities discovered by Löwner, Cordes, Jensen-Hansen, McIntosh, Heinz-Kato, and Reid, respectively. The next result is to show that Proposition is indeed a special case of Theorem 2.

Corollary. *Let $A, B, C, D \geq O$, T and K be arbitrary operators with AK selfadjoint, and let E be any contraction operator. Then, for $\alpha, \beta \in (0, 1]$, $r, s \geq 0$, $p, q \geq 1$, $t \in [1, 2]$ and $\delta \geq 1$, each inequality in Theorem 2 implies corresponding inequality in Proposition.*

Proof. The proof follows immediately by letting $r = s = 0$ in every inequality of Theorem 2. Therefore, we obtain corresponding inequalities in Proposition which is denoted by the same number in the bracket, i.e., from (1) through (11′), (14) and (15). As for others, if

$\beta = 1 - \alpha$ in (12) of Theorem 2, then we get (12) in Proposition; and if $\alpha = \beta = \frac{1}{2}$ in (15) of Theorem 2, then we have (15′) in Proposition. Finally, (13) in Theorem 2 becomes

$$|(|T^*|^\beta U|T|^\alpha x, y)| \leq \|A^\alpha x\| \|B^\beta y\|$$

if $r = s = 0$. But $|T^*|^\beta U|T|^\alpha = U|T|^{\alpha+\beta}$ by Lemma, which is equal to $T|T|^{\alpha+\beta-1}$ as $T = U|T|$. Hence, we have (13) in Proposition. This completes the proof.

4. Characteristic Property of the Heinz-Kato Inequality

In this finally section we shall characterize the Heinz-Kato inequality, i.e., (12) in Proposition, without polar decomposition of operator involved. In fact, we use a set of vectors in the range of a positive operator and a set of vectors in H recursively defined (cf. [11]) to construct some inequalities which are all equivalent to the Heinz-Kato inequality. Moreover, we see from the theorem below that each inequality shows sharper form of the Heinz-Kato inequality.

Theorem 3. *Let $A, B \geq O$ and T be arbitrary. If $T^*T \leq A^2$, $TT^* \leq B^2$, and $\alpha \in [0, 1]$, then the following are equivalent for all $x, y \in$H.*

(1) $|(Tx, y)| \leq \|A^\alpha x\| \|B^{1-\alpha} y\|$ *(Heinz-Kato inequality [10]);*

(2) $\|B^{1-\alpha} y\|^2 \sum_1^n \frac{|(u_{i-1}, A^{2\alpha} z_i)|^2}{\|A^\alpha z_i\|^2} \leq \|A^\alpha x\|^2 \|B^{1-\alpha} y\|^2 - |(Tx, y)|^2$, *where $\{A^\alpha \cdot z_i\}_1^n$ is a set of nonzero vectors, T^*y and $\{z_i\}_1^n$ are orthogonal, and $\{u_i\}_1^n$ is a sequence of vectors recursively defined by $u_i = u_{i-1} - \frac{(u_{i-1}, A^{2\alpha} z_i) z_i}{\|A^\alpha z_i\|^2}$, $i = 1, \cdots, n$, and $u_0 = x$;*

(3) $\|B^{1-\alpha} y\|^2 \frac{|(x, A^{2\alpha} z)|^2}{\|A^\alpha z\|^2} \leq \|A^\alpha x\|^2 \|B^{1-\alpha} y\|^2 - |(Tx, y)|^2$, *where $A^\alpha z \neq 0$, and T^*y and z are orthogonal;*

(4) $\|B^{1-\alpha} y\|^2 \sum_1^n |(x, A^{2\alpha} w_i)|^2 \leq \|A^\alpha x\|^2 \|B^{1-\alpha} y\|^2 - |(Tx, y)|^2$, *where $\{A^\alpha \cdot w_i\}_1^n$ is a set of orthonormal vectors, and T^*y and $\{w_i\}_1^n$ are orthogonal;*

(5) $\|B^{1-\alpha} y\|^2 |(x, A^{2\alpha} w)|^2 \leq \|A^\alpha x\|^2 \|B^{1-\alpha} y\|^2 - |(Tx, y)|^2$, *where $A^\alpha w$ is a unit vector, and T^*y and w are orthogonal.*

Proof. (1)⇒(2). For $i = 1, \cdots, n$, we have

$$(u_i, A^{2\alpha} z_i) = (u_{i-1}, A^{2\alpha} z_i) - \frac{(u_{i-1}, A^{2\alpha} z_i)}{\|A^\alpha z_i\|^2} (z_i, A^{2\alpha} z_i) = 0$$

by the definition of u_i. It follows that

$$(A^{2\alpha} u_{i-1}, u_{i-1}) = (A^{2\alpha} u_i, u_{i-1}) + \frac{(u_{i-1}, A^{2\alpha} z_i)}{\|A^\alpha z_i\|^2} (A^{2\alpha} z_i, u_{i-1});$$

and

$$\begin{aligned} \|A^\alpha u_{i-1}\|^2 &= \left(A^{2\alpha} u_i, u_i + \frac{(u_{i-1}, A^{2\alpha} z_i) z_i}{\|A^\alpha z_i\|^2} \right) \\ &+ \frac{(u_{i-1}, A^{2\alpha} z_i)}{\|A^\alpha z_i\|^2} \left(A^{2\alpha} z_i, u_i + \frac{(u_{i-1}, A^{2\alpha} z_i) z_i}{\|A^\alpha z_i\|^2} \right) \\ &= \|A^\alpha u_i\|^2 + \frac{|(u_{i-1}, A^{2\alpha} z_i)|^2}{\|A^\alpha z_i\|^2}. \end{aligned}$$

But

$$(Tu_{i-1}, y) = (Tu_i, y) + \frac{(u_{i-1}, A^{2\alpha}z_i)}{\|A^\alpha z_i\|^2}(Tz_i, y) = (Tu_i, y)$$

since T^*y and $\{z_i\}_1^n$ are orthogonal by assumption. Therefore,

$$\begin{aligned}&\|A^\alpha u_{i-1}\|^2\|B^{1-\alpha}y\|^2 - |(Tu_{i-1}, y)|^2 \\ &= \|B^{1-\alpha}y\|^2\left[\|A^\alpha u_i\|^2 + \frac{|(u_{i-1}, A^{2\alpha}z_i)|^2}{\|A^\alpha z_i\|^2}\right] - |(Tu_i, y)|^2.\end{aligned}$$

In the identity, above let $i = 1, \cdots, n$ and then add up the n equations so obtained we find that

$$\begin{aligned}&\|A^\alpha x\|^2\|B^{1-\alpha}y\|^2 - |(Tx, y)|^2 \\ &= \|B^{1-\alpha}y\|^2 \sum_1^n \frac{|(u_{i-1}, A^{2\alpha}z_i)|^2}{\|A^\alpha z_i\|^2} + \|A^\alpha u_n\|^2\|B^{1-\alpha}y\|^2 - |(Tu_n, y)|^2 \\ &\geq \|B^{1-\alpha}y\|^2 \sum_1^n \frac{|(u_{i-1}, A^{2\alpha}z_i)|^2}{\|A^\alpha z_i\|^2}\end{aligned}$$

since $\|A^\alpha u_n\|^2\|B^{1-\alpha}y\|^2 \geq |(Tu_n, y)|^2$ by (1).

(2)⇒(3). This is immediate from (2) by letting $n = 1$ and $z_1 = z$.

(2)⇒(4). This is a special case of (2), and the definition of u_i becomes $u_i = u_{i-1} - (u_{i-1}, A^{2\alpha}w_i)w_i$ for $i = 1, \cdots, n$ with $u_0 = x$. Now, $u_1 = x - (x, A^{2\alpha}w_1)w_1$ and so

$$(u_1, A^{2\alpha}w_2) = (x, A^{2\alpha}w_2) - (x, A^{2\alpha}w_1)(w_1, A^{2\alpha}w_2) = (x, A^{2\alpha}w_2).$$

By induction on n, we get

$$(u_{i-1}, A^{2\alpha}w_i) = (x, A^{2\alpha}w_i), \quad i = 1, \cdots, n.$$

(4)⇒(5). In (4) let $n = 1$ and $w_1 = w$. The set consisting of the unit vector $A^\alpha w$ only is orthonormal.

As the left side of each inequality (3) and (5) is positive, each inequality implies (1), and this completes the proof.

In conclusion, it is hoped that results obtained in this paper may be of use in investigations of the Furuta inequality and its applications (especially equivalence relations), and operator inequalities in general.

References

1. H. O. Cordes, *Spectral Theory of Linear Differential Operators and Comparison Algebras*, London Mathematical Society Lecture Note Series, vol. 76, 1987.
2. M. Fujii and T. Furuta, *Löwner-Heinz, Cordes and Heinz-Kato inequalities*, Math. Japon. **38** (1993), 73–78.
3. M. Fujii, T. Furuta and R. Nakamoto, *Norm inequalities in the Corach-Porta-Recht theory and operator means*, Illinois J. Math. **40** (1996), 527–534.
4. T. Furuta, *$A \geq B \geq O$ assures $(B^rA^pB^r)^{1/q} \geq B^{(p+2r)/q}$ for $r \geq 0$, $p \geq 0$, $q \geq 1$ with $(1+2r)q \geq p+2r$*, Proc. Amer. Math. Soc. **101** (1987), 85–88.
5. ———, *Norm inequalities equivalent to Löwner-Heinz theorem*, Rev. Math. Phys. **1** (1989), 135–137.
6. ———, *Generalization of Heinz-Kato theorem via Furuta inequality*, Operator Theory **62** (1993), 77–83.
7. ———, *Invitation To Linear Operators*, Taylor & Francis, London and New York, 2001.
8. F. Hansen, *An operator inequality*, Math. Ann. **246** (1980), 249–250.

9. E. Heinz, *Beiträge zur Störungstheorie der Spektralzerlegung*, Math. Ann. **123** (1951), 415–538.
10. T. Kato, *Notes on some inequalities for linear operators*, Math. Ann. **125** (1952), 208–212.
11. C.-S. Lin, *Heinz's inequality and Bernstein's inequality*, Proc. Amer. Math. Soc. **125** (1997), 2319–2325.
12. ———, *On Heinz-Kato type characterizations of the Furuta inequality*, Nihonkai Math. J. **9** (1998), 187–191.
13. ———, *Inequalities of Reid type and Furuta*, Proc. Amer. Math. Soc. **129** (2001), 855–859.
14. ———, *On Heinz-Kato type characterizations of the Furuta inequality II*, Math. Inequalities Appli. **2** (1999), 283–287.
15. K. Löwner, *Über monotone Matrixfunktionen*, Math. Z. **38** (1934), 177–216.
16. A. McIntosh, *Heinz inequalities and perturbation of spectral falilies*, Macquarie Math. Reports, 1979.
17. W. T. Reid, *Symmetrizable completely continuous linear transformations in Hilbert space*, Duke Math. J. **18** (1951), 41–56.

In: Inequality Theory and Applications, Volume 4
Editors: Yeol Je Cho et al., pp.93–109

ISBN 1-59454-874-9

IMPROVED BOUNDS ON ENTROPY MEASURES FOR MIXED POPULATIONS

M. MATIĆ[1], C. E. M. PEARCE[2] AND J. PEČARIĆ[3]

ABSTRACT. We derive a number of bounds relating entropy measures for a mixed population to parameters of the n constituent populations. The results extend some recent studies of Dragomir and Goh included in [3] for the two-population case and are tighter in the $n = 2$ case. It refines and extends an earlier study by the authors [6].

1. Introduction

Recently a number of improvements have been made to some of the fundamental inequalities for various entropy measures to do with discrete-valued random variables in information theory. We note in particular the work of Dragomir and Goh [1]-[3]. See also the present authors [4]-[7].

In this article we address the situation of n random variables X_i having the same finite range but different probability distributions. If we mix populations represented by these random variables with respective proportions α_i (with $\alpha_i > 0$ and $\sum_{i=1}^{n} \alpha_i = 1$), we derive a new population represented by the random variable X with

$$p(x) = \sum_{i=1}^{n} \alpha_i p_i(x), \tag{1.1}$$

where $p_i(x) = P(X_i = x)$ gives the probability distribution of X_i $(i = 1, \dots, n)$ and $p(x)$ that of X. Such a situation can arise, in particular, in problems involving a number of sources being fed into a common channel.

We are interested in relations between information measures for X and those for the original random variables X_i. For example, the entropy of X is defined by

$$H(X) := \sum_{x} p(x) \log(1/p(x)).$$

Received April 18, 2003.
2000 *Mathematics Subject Classification.* 94A17, 26D15.
Key words and phrases. Bounds, entropy measures, mixed populations.
[1] FESB, Mathematics Department, University of Split, R. Boškovića bb, 21000 Split, Croatia. E-mail: mmatic@fesb.hr
[2] School of Applied Mathematics, University of Adelaide, Adelaide SA 5005, Australia. E-mail:l cpearce@maths.adelaide.edu.au
[3] Faculty of Textile Technology, University of Zagreb, Pierottijeva 6, Zagreb, Croatia. E-mail: pecaric@hazu.hr

Here log denotes a logarithm taken to a fixed base $b > 1$. We assume this notation in the remainder of the paper without further comment. The entropy function H is a concave functional of the probability distribution, that is,

$$\sum_{i=1}^{n} \alpha_i H(X_i) \leq H(X), \tag{1.2}$$

so that $H(X) - \sum_i \alpha_i H(X_i)$ is bounded below by zero. But what can be said about upper bounds for this quantity? We provide several answers to this question in Section 2, along with a simple derivation of (1.2).

In Sections 3, 4 and 5 we give corresponding results for joint entropy, conditional entropy and mutual information respectively in the context of mixed populations. Our results both extend known results for the case $n = 2$ derived in [3] to general n and sharpen existing results for the case $n = 2$.

These questions were addressed in [6], which is a precursor to the present chapter. Here we make use of refinements to the tools in [6] to give stronger versions of the results therein.

Our arguments make use of several basic results. One is the following lemma established as [5, Lemma 1.]. The authors have found this to have a key role in the study of entropy measures.

Lemma A. (i) *Suppose* $\xi_k > 0$ *and* $p_k > 0$ $(k = 1, \cdots, n)$ *with* $\sum_{k=1}^{n} p_k = 1$. *Then*

$$0 \leq \log\left(\sum_{k=1}^{n} p_k \xi_k\right) - \sum_{k=1}^{n} p_k \log \xi_k \leq \log\left(\sum_{k=1}^{n} p_k \xi_k \sum_{j=1}^{n} \frac{p_j}{\xi_j}\right).$$

(ii) *Also, if* $\rho := \max_{i,k}\{\xi_i/\xi_k\}$, *then*

$$0 \leq \log\left(\sum_{k=1}^{n} p_k \xi_k\right) - \sum_{k=1}^{n} p_k \log \xi_k \leq \log\left[\frac{1}{4}\left(\sqrt{\rho} + \frac{1}{\sqrt{\rho}}\right)^2\right].$$

We shall also need the following discrete version of the Grüss inequality (see [9, Chapter 10]).

Lemma B. *Let* (a_k) *and* (b_k) *be* n*-tuples of real numbers and* $p_k \geq 0$ $(1 \leq k \leq n)$. *If* $\alpha \leq a_k \leq A$ *and* $\beta \leq b_k \leq B$, *then*

$$\left|\sum_{k=1}^{n} p_k \sum_{j=1}^{n} p_j a_j b_j - \sum_{k=1}^{n} p_k a_k \sum_{j=1}^{n} p_j b_j\right| \leq \frac{1}{4}(A - \alpha)(B - \beta)\left(\sum_{k=1}^{n} p_k\right)^2.$$

Some of our arguments depend on the elementary inequality

$$\log x \leq \frac{1}{\ln b}(x - 1), \tag{1.3}$$

which is valid for all $x > 0$. Also we use the well–known inequality

$$\prod_{i=1}^{n} z_i^{\alpha_i} \leq \sum_{i=1}^{n} \alpha_i z_i \tag{1.4}$$

between geometric and arithmetic means This holds whenever $z_i > 0$, $\alpha_i > 0$ $(i = 1, \cdots, n)$ and $\sum_{i=1}^{n} \alpha_i = 1$.

2. Entropy for a Mixed Population

In this section we suppose as in the introduction that populations given by $p_i(\cdot)$ are mixed in accordance with probability weights α_i $(1 \le i \le n)$ to produce a population given by $p(\cdot)$, that is, (1.1) holds. We derive several upper bounds for $H(X) - \sum_{i=1}^n \alpha_i H(X_i)$. We show *en passant* that a lower bound is zero, that is to say, H is concave.

Denote by R the common range of X_i $(i = 1, \cdots, n)$ and X. Without loss of generality we may assume that for each $x \in R$, $p_i(x) > 0$ holds for at least one value of i. Since $\alpha_i > 0$, we have that $p(x) > 0$ for each $x \in R$. However it may happen that $p_i(x) = 0$ for some indices i and for some $x \in R$. Let us define the sets U_i and the values S_i by

$$U_i := \{x \in R : p_i(x) > 0\}, \; S_i := \sum\nolimits_{x \in U_i} p(x), \qquad (i = 1, \cdots, n).$$

It is evident that $\sum_{x \in U_i} p_i(x) = 1$ and $0 < S_i \le 1$, for all $i = 1, \cdots, n$, but we may have $S_i < 1$ for some indices i.

Theorem 2.1. *With X_i and X as above, we have*

$$\begin{aligned} -\log S \le H(X) - \sum_{i=1}^n \alpha_i H(X_i) &\le \sum_{i=1}^n \alpha_i \log\left[\sum_{x \in U_i} \frac{p_i^2(x)}{p(x)}\right] \\ &= \log\left[\prod_{i=1}^n \left(\sum_{x \in U_i} \frac{p_i^2(x)}{p(x)}\right)^{\alpha_i}\right] \le \frac{1}{\ln b}\left[\prod_{i=1}^n \left(\sum_{x \in U_i} \frac{p_i^2(x)}{p(x)}\right)^{\alpha_i} - 1\right] \\ &\le \frac{1}{\ln b}\left[\sum_{x \in R}\sum_{i=1}^n \alpha_i \frac{p_i^2(x)}{p(x)} - 1\right], \end{aligned} \tag{2.1}$$

where $S = \prod_{i=1}^n S_i^{\alpha_i}$.

Proof. By (1.1) we have

$$\begin{aligned} &H(X) - \sum_{i=1}^n \alpha_i H(X_i) \\ &= \sum_{x \in R} p(x) \log \frac{1}{p(x)} - \sum_{i=1}^n \alpha_i \sum_{x \in U_i} p_i(x) \log \frac{1}{p_i(x)} \\ &= \sum_{x \in R}\sum_{i=1}^n \alpha_i p_i(x) \log \frac{1}{p(x)} - \sum_{i=1}^n \alpha_i \sum_{x \in U_i} p_i(x) \log \frac{1}{p_i(x)} \\ &= -\sum_{i=1}^n \alpha_i \sum_{x \in U_i} p_i(x) \log \frac{p(x)}{p_i(x)}. \end{aligned} \tag{2.2}$$

The last equality in (2.2) follows from the fact that $p_i(x)\log(1/p(x)) = 0$ when $x \in R \setminus U_i$. From Lemma A (i) with $p_k = p_i(x)$ and $\xi_k = p(x)/p_i(x)$ $(x \in U_i)$, we get for fixed i

$$\begin{aligned} 0 &\le \log\left[\sum_{x \in U_i} p_i(x) \frac{p(x)}{p_i(x)}\right] - \sum_{x \in U_i} p_i(x) \log \frac{p(x)}{p_i(x)} \\ &= \log S_i - \sum_{x \in U_i} p_i(x) \log \frac{p(x)}{p_i(x)} \\ &\le \log\left[\sum_{y \in U_i} p_i(y) \frac{p(y)}{p_i(y)} \sum_{x \in U_i} p_i(x) \frac{p_i(x)}{p(x)}\right] \\ &= \log S_i + \log\left[\sum_{x \in U_i} \frac{p_i^2(x)}{p(x)}\right], \qquad i = 1, \cdots, n. \end{aligned}$$

This can be rewritten as

$$-\log S_i \le -\sum_{x\in U_i} p_i(x)\log\frac{p(x)}{p_i(x)} \le \log\left[\sum_{x\in U_i}\frac{p_i^2(x)}{p(x)}\right], \quad i=1,\cdots,n. \tag{2.3}$$

Multiplying (2.3) by $\alpha_i > 0$ and summing over $i = 1,\cdots,n$ yields

$$-\sum_{i=1}^{n}\alpha_i\log S_i \le -\sum_{i=1}^{n}\alpha_I\sum_{x\in U_i}p_i(x)\log\frac{p(x)}{p_i(x)} \le \sum_{i=1}^{n}\alpha_I\log\left[\sum_{x\in U_i}\frac{p_i^2(x)}{p(x)}\right].$$

By (2.2) and the fact that $\log S = \sum_{i=1}^{n}\alpha_i\log S_i$ these are the first and second inequalities in (2.1). The third follows by (1.3), while the last is a consequence of (1.4).

Remark. Note that $0 < S \le 1$ implies $-\log S \ge 0$. Thus the inequality (1.2) is a consequence of the first inequality in (2.1).

Theorem 2.2. *Define*

$$M_i := \max_{x\in U_i}\{p_i(x)/p(x)\}, \; m_i := \min_{x\in U_i}\{p_i(x)/p(x)\} \quad (i=1,\cdots,n).$$

Then

$$\begin{aligned} 0 &\le \log S + H(X) - \sum_{i=1}^{n}\alpha_i H(X_i) \le \sum_{i=1}^{n}\alpha_i\log\left[\frac{1}{4}(M_i-m_i)^2S_i^2+1\right] \\ &= \log\left\{\prod_{i=1}^{n}\left[\frac{1}{4}(M_i-m_i)^2S_i^2+1\right]^{\alpha_i}\right\} \\ &\le \frac{1}{\ln b}\left\{\prod_{i=1}^{n}\left[\frac{1}{4}(M_i-m_i)^2S_i^2+1\right]^{\alpha_i}-1\right\} \\ &\le \frac{1}{4\ln b}\sum_{i=1}^{n}\alpha_i(M_i-m_i)^2S_i^2. \end{aligned} \tag{2.4}$$

Also

$$\begin{aligned} 0 &\le \log S + H(X) - \sum_{i=1}^{n}\alpha_i H(X_i) \le \sum_{i=1}^{n}\alpha_i\log\left[\frac{1}{4}\left(\sqrt{\rho_i}+\frac{1}{\sqrt{\rho_i}}\right)^2\right] \\ &= \log\left[\frac{1}{4}\prod_{i=1}^{n}\left(\sqrt{\rho_i}+\frac{1}{\sqrt{\rho_i}}\right)^{2\alpha_i}\right] \le \frac{1}{\ln b}\left[\frac{1}{4}\prod_{i=1}^{n}\left(\sqrt{\rho_i}+\frac{1}{\sqrt{\rho_i}}\right)^{2\alpha_i}-1\right] \\ &\le \frac{1}{4\ln b}\sum_{i=1}^{n}\alpha_i\left(\sqrt{\rho_i}-\frac{1}{\sqrt{\rho_i}}\right)^2, \end{aligned} \tag{2.5}$$

where $\rho_i = M_i/m_i$ $(i=1,\cdots,n)$.

Proof. By Lemma B with $p_k = p(x)$ and $a_k = b_k = p_i(x)/p(x)$ $(x\in U_i)$, we have for fixed i that

$$\begin{aligned} S_i\sum_{x\in U_i}\frac{p_i^2(x)}{p(x)}-1 &= \sum_{y\in U_i}p(y)\sum_{x\in U_i}p(x)\left[\frac{p_i(x)}{p(x)}\right]^2-\left[\sum_{x\in U_i}p(x)\frac{p_i(x)}{p(x)}\right]^2 \\ &\le \frac{1}{4}(M_i-m_i)^2S_i^2, \quad i=1,\cdots,n. \end{aligned}$$

Therefore

$$\log S_i + \log\left[\sum_{x\in U_i} \frac{p_i^2(x)}{p(x)}\right] \le \log\left[\frac{1}{4}(M_i - m_i)^2 S_i^2 + 1\right], \; i = 1, \cdots, n.$$

Multiplying this by $\alpha_i > 0$ and summing over $i = 1, \cdots, n$ produces

$$\log S + \sum_{i=1}^{n} \alpha_i \log\left[\sum_{x\in U_i} \frac{p_i^2(x)}{p(x)}\right] \le \sum_{i=1}^{n} \alpha_i \log\left[\frac{1}{4}(M_i - m_i)^2 S_i^2 + 1\right].$$

In combination with the first and second inequalities in (2.1), this gives the first and second inequalities in (2.4). The third and fourth follows by (1.3) and (1.4), respectively. Further, for fixed i, set $p_k = p_i(x)$ and $\xi_k = p(x)/p_i(x)$ $(x \in U_i)$ in Lemma A (ii) and note that

$$\rho_i = \max_{j,k}\{\xi_j/\xi_k\} = (1/m_i)/(1/M_i) = M_i/m_i.$$

We derive

$$\begin{aligned} 0 &\le \log\left[\sum_{x\in U_i} p_i(x)\frac{p(x)}{p_i(x)}\right] - \sum_{x\in U_i} p_i(x) \log \frac{p(x)}{p_i(x)} \\ &= \log S_i - \sum_{x\in U_i} p_i(x) \log \frac{p(x)}{p_i(x)} \\ &\le \log\left[\frac{1}{4}\left(\sqrt{\rho_i} + \frac{1}{\sqrt{\rho_i}}\right)^2\right], \qquad i = 1, \cdots, n. \end{aligned}$$

Multiplying this by $\alpha_i > 0$ and summing over $i = 1, \cdots, n$ yields

$$0 \le \log S - \sum_{i=1}^{n} \alpha_i \sum_{x\in U_i} p_i(x) \log \frac{p(x)}{p_i(x)} \le \sum_{i=1}^{n} \alpha_i \log\left[\frac{1}{4}\left(\sqrt{\rho_i} + \frac{1}{\sqrt{\rho_i}}\right)^2\right]. \tag{2.6}$$

Combining (2.6) and (2.2) gives the first and second inequalities in (2.5). The third and fourth follow from (1.3) and (1.4), respectively.

Remark. Let $\max_{1\le i\le n}\{M_i - m_i\} = M - m$. From (2.4) we obtain

$$\begin{aligned} 0 &\le \log S + H(X) - \sum_{i=1}^{n} \alpha_i H(X_i) \le \log\left\{\prod_{i=1}^{n}\left[\frac{1}{4}(M - m)^2 S_i^2 + 1\right]^{\alpha_i}\right\} \\ &\le \frac{(M-m)^2}{4\ln b} \sum_{i=1}^{n} \alpha_i S_i^2, \end{aligned}$$

where the third inequality follows from (1.3) and (1.4). If in addition $p_i(x) > 0$ for all indices i and all $x \in R$, then $S_i = 1$ for all i and these inequalities reduce to

$$0 \le H(X) - \sum_{i=1}^{n} \alpha_i H(X_i) \le \log\left[\frac{1}{4}(M - m)^2 + 1\right] \le \frac{1}{4\ln b}(M - m)^2. \tag{2.7}$$

Furthermore, the function $f(x) = (\sqrt{x} + 1/\sqrt{x})^2$ is increasing for $x \ge 1$. Using (1.3) and (1.4) we derive from (2.5) that

$$0 \le \log S + H(X) - \sum_{i=1}^{n} \alpha_i H(X_i) \le \log\left[\frac{1}{4}\left(\sqrt{\rho} + \frac{1}{\sqrt{\rho}}\right)^2\right] \le \frac{1}{4\ln b}\left(\sqrt{\rho} - \frac{1}{\sqrt{\rho}}\right)^2,$$

where $\rho = \max_{1\le i\le n}\{\rho_i\}$.

Corollary 2.3. *Suppose that $p_i(x) > 0$ for all $i = 1, \cdots, n$ and for all $x \in R$. If*

$$\max_{x,y\in R}\left\{\max_{1\le i\le n}\left\{\left|\frac{p_i(x)}{p(x)} - \frac{p_i(y)}{p(y)}\right|\right\}\right\} \le 2\sqrt{b^\varepsilon - 1}$$

for some $\varepsilon = \varepsilon(\alpha_1, \cdots, \alpha_n) > 0$, then

$$0 \le H(X) - \sum_{i=1}^{n} \alpha_i H(X_i) \le \varepsilon.$$

Proof. Note that

$$\max_{x,y\in R}\left\{\max_{1\le i\le n}\left\{\left|\frac{p_i(x)}{p(x)} - \frac{p_i(y)}{p(y)}\right|\right\}\right\} = \max_{1\le i\le n}\{M_i - m_i\} = M - m.$$

The result follows from the second inequality in (2.7).

Remark. In the case $n = 2$, this strengthens a result of Dragomir and Goh [3, Cor. 4.2], which gives the result with $\sqrt{2\varepsilon \ln b}$ in place of $2\sqrt{b^\varepsilon - 1}$. Note that $2\sqrt{b^\varepsilon - 1} > 2\sqrt{\varepsilon \ln b} > \sqrt{2\varepsilon \ln b}$.

The upper bounds on $H(X) - \sum_{i=1}^{n} \alpha_i H(X_i)$ derived in Theorems 2.1 and 2.2 involve the quantities α_i. To conclude this section we derive one that does not. We employ the following result.

Lemma 2.4. *Suppose*

$$g(\alpha_1, \ldots, \alpha_n) := \frac{\sum_{i=1}^{n} \alpha_i t_i u_i^2}{\sum_{i=1}^{n} \alpha_i t_i u_i} \quad (0 \le \alpha_i \le 1),$$

where $0 \le t_i, u_i \le 1$ $(1 \le i \le n)$ are fixed and

$$I := \{i : \alpha_i t_i u_i > 0\} \ne \emptyset.$$

Then

$$\min_i\{u_i\} \le \min_{i\in I}\{u_i\} \le g \le \max_{i\in I}\{u_i\} \le \max_i\{u_i\} \quad \forall \alpha_i \in [0, 1].$$

Proof. We may regard g as an average $g = \sum_{i=1}^{n} w_i u_i$ of the values u_i with the weights

$$w_i := \alpha_i t_i u_i \Big/ \sum_j \alpha_j t_j u_j.$$

Since $w_i \ge 0$ and $\sum_i w_i = 1$, the quantity g lies between the least and greatest of the values u_i for which $i \in I$. The desired result follows.

Theorem 2.5. *We have*

$$-\log S \le H(X) - \sum_{i=1}^{n} \alpha_i H(X_i) \le \frac{1}{\ln b}\left[\sum_{x\in R} \max_{1\le i\le n}\{p_i(x)\} - 1\right].$$

Proof. Since $\alpha_i > 0$ for each i and for each $x \in R$ we have $p_i(x) > 0$ for at least one $i \in \{1, 2, \ldots, n\}$, we may set $t_i = 1$, $u_i = p_i(x)$ in Lemma 2.4 to obtain

$$\sum_{i=1}^{n} \alpha_i p_i(x)^2 / p(x) \le \max_{1\le i\le n}\{p_i(x)\}.$$

The desired result follows from Theorem 2.1.

3. Joint Entropy for a Mixed Population

Now we shall give results for joint entropy analogous to those for entropy. For X, Y discrete random variables with finite ranges, the joint entropy of X and Y is defined by

$$H(X,Y) := \sum_{x,y} p(x,y)\log(1/p(x,y))$$

(see [8, p. 25]), where $p(x,y) := P\{X = x, Y = y\}$ and the summation is over those pairs (x,y) for which $p(x,y) > 0$.

Theorem 3.1. *Let (X_i, Y_i) $(1 \le i \le n)$ be a pairs of discrete random variables with the same finite range R and respective joint probability distributions $p_i(x,y)$. Suppose that the pair (X,Y) of discrete random variables satisfies*

$$p(x,y) := \sum_{i=1}^{n} \alpha_i p_i(x,y), \quad \alpha_i > 0, \ \sum_{i=1}^{n} \alpha_i = 1.$$

Suppose further that $p(x,y) > 0$ for all $(x,y) \in R$ and define

$$W_i := \{(x,y) \in R : p_i(x,y) > 0\}, \ S_i := \sum\nolimits_{(x,y)\in W_i} p(x,y), \quad (i = 1, \cdots, n).$$

Then

$$\begin{aligned}
-\log S \le H(X,Y) - \sum_{i=1}^{n} \alpha_i H(X_i,Y_i) &\le \sum_{i=1}^{n} \alpha_i \log\left[\sum_{(x,y)\in W_i} \frac{p_i^2(x,y)}{p(x,y)}\right] \\
= \log\left[\prod_{i=1}^{n}\left(\sum_{(x,y)\in W_i} \frac{p_i^2(x,y)}{p(x,y)}\right)^{\alpha_i}\right] &\le \frac{1}{\ln b}\left[\prod_{i=1}^{n}\left(\sum_{(x,y)\in W_i} \frac{p_i^2(x,y)}{p(x,y)}\right)^{\alpha_i} - 1\right] \\
\le \frac{1}{\ln b}\left[\sum_{(x,y)\in R}\sum_{i=1}^{n} \alpha_i \frac{p_i^2(x,y)}{p(x,y)} - 1\right],
\end{aligned}$$

where $S = \prod_{i=1}^{n} S_i^{\alpha_i}$.

Proof. We argue as in Theorem 2.1 with $p(x,y), p_i(x,y)$ in place of $p(x), p_i(x)$ respectively. □

Theorem 3.2. *Let (X_i, Y_i) $(1 \le i \le n)$ and (X,Y) be as in the previous theorem. For $i = 1, \cdots, n$, define*

$$M_i := \max_{(x,y)\in W_i} \{p_i(x,y)/p(x,y)\}, \ m_i := \min_{(x,y)\in W_i} \{p_i(x,y)/p(x,y)\}.$$

Then

$$\begin{aligned}
0 \le \log S + H(X,Y) - \sum_{i=1}^{n} \alpha_i H(X_i,Y_i) &\le \sum_{i=1}^{n} \alpha_i \log\left[\frac{1}{4}(M_i - m_i)^2 S_i^2 + 1\right] \\
= \log\left\{\prod_{i=1}^{n}\left[\frac{1}{4}(M_i - m_i)^2 S_i^2 + 1\right]^{\alpha_i}\right\} &\le \frac{1}{\ln b}\left\{\prod_{i=1}^{n}\left[\frac{1}{4}(M_i - m_i)^2 S_i^2 + 1\right]^{\alpha_i} - 1\right\} \\
\le \frac{1}{4\ln b}\sum_{i=1}^{n} \alpha_i (M_i - m_i)^2 S_i^2.
\end{aligned}$$

Also we have

$$0 \le \log S + H(X,Y) - \sum_{i=1}^{n} \alpha_i H(X_i,Y_i) \le \sum_{i=1}^{n} \alpha_i \log\left[\frac{1}{4}\left(\sqrt{\rho_i} + \frac{1}{\sqrt{\rho_i}}\right)^2\right]$$
$$= \log\left[\frac{1}{4}\prod_{i=1}^{n}\left(\sqrt{\rho_i} + \frac{1}{\sqrt{\rho_i}}\right)^{2\alpha_i}\right] \le \frac{1}{\ln b}\left[\frac{1}{4}\prod_{i=1}^{n}\left(\sqrt{\rho_i} + \frac{1}{\sqrt{\rho_i}}\right)^{2\alpha_i} - 1\right]$$
$$\le \frac{1}{4\ln b}\sum_{i=1}^{n}\alpha_i\left(\sqrt{\rho_i} - \frac{1}{\sqrt{\rho_i}}\right)^2,$$

where $\rho_i = M_i/m_i$ $(i = 1, \cdots, n)$.

Proof. The argument follows that of Theorem 2.2.

Again we may argue as in the example following Theorem 2.2 to derive the following.

Corollary 3.3. *Suppose that* $p_i(x,y) > 0$ *for all* $i = 1, \cdots, n$ *and for all* $(x,y) \in R$. *If*

$$\max_{(x,y),(u,v)\in R}\left\{\max_{1\le i\le n}\left\{\left|\frac{p_i(x,y)}{p(x,y)} - \frac{p_i(u,v)}{p(u,v)}\right|\right\}\right\} \le 2\sqrt{b^{\varepsilon} - 1}$$

for some $\varepsilon = \varepsilon(\alpha_1, \cdots, \alpha_n) > 0$, *then*

$$0 \le H(X,Y) - \sum_{i=1}^{n} \alpha_i H(X_i,Y_i) \le \varepsilon.$$

Finally we have a result analogous to Theorem 2.5.

Theorem 3.4. *Let* (X_i, Y_i) *and* (X, Y) *be as in Theorem 3.1. Then*

$$-\log S \le H(X,Y) - \sum_{i=1}^{n} \alpha_i H(X_i,Y_i) \le \frac{1}{\ln b}\left[\sum_{(x,y)\in R} \max_{1\le i\le n}\{p_i(x,y)\} - 1\right].$$

4. Conditional Entropy for a Mixed Population

For a pair of discrete random variables X, Y with finite ranges, the conditional entropy of X given Y is defined by

$$H(X|Y) := \sum_{x,y} p(x,y)\log(1/p(x|y)),$$

where $p(x,y) := P\{X = x, Y = y\}$, $p(x|y) := P\{X = x|Y = y\} = p(x,y)/p(y)$ and the summation is over those pairs (x,y) for which $p(x,y) > 0$ (see [8, p. 22]).

In this section we prove that conditional entropy is a concave functional of the joint probability distribution $p(x,y)$. Also we derive some upper bounds for the difference $H(X|Y) - \sum_{i=1}^{n} \alpha_i H(X_i|Y_i)$.

Theorem 4.1. *Suppose that* (X_i, Y_i) $(1 \le i \le n)$ *and* (X, Y) *are as in Theorem 3.1. Also, with* W_i $(1 \le i \le n)$ *as in Theorem 3.1, define*

$$T_i := \sum\nolimits_{(x,y)\in W_i} p_i(y)p(x|y), \quad (i = 1, \cdots, n).$$

Then we have

$$
\begin{aligned}
-\log T \le H(X|Y) - \sum_{i=1}^{n} \alpha_i H(X_i|Y_i) &\le \sum_{i=1}^{n} \alpha_i \log\left[\sum_{(x,y)\in W_i} \frac{p_i(y)p_i^2(x|y)}{p(x|y)}\right] \\
&= \log\left[\prod_{i=1}^{n}\left(\sum_{(x,y)\in W_i} \frac{p_i(y)p_i^2(x|y)}{p(x|y)}\right)^{\alpha_i}\right] \\
&\le \frac{1}{\ln b}\left[\prod_{i=1}^{n}\left(\sum_{(x,y)\in W_i} \frac{p_i(y)p_i^2(x|y)}{p(x|y)}\right)^{\alpha_i} - 1\right] \\
&\le \frac{1}{\ln b}\left[\sum_{(x,y)\in R}\sum_{i=1}^{n} \alpha_i \frac{p_i(y)p_i^2(x|y)}{p(x|y)} - 1\right],
\end{aligned}
\tag{4.1}
$$

where $T = \prod_{i=1}^{n} T_i^{\alpha_i}$.

Proof. Since $p(x,y) = \sum_{i=1}^{n} \alpha_i p_i(x,y)$, we have by the definition of conditional entropy that

$$
\begin{aligned}
&H(X|Y) - \sum_{i=1}^{n} \alpha_i H(X_i|Y_i) \\
&= \sum_{(x,y)\in R}\left(\sum_{i=1}^{n} \alpha_i p_i(x,y)\right)\log\frac{1}{p(x|y)} - \sum_{i=1}^{n}\alpha_i \sum_{(x,y)\in W_i} p_i(x,y)\log\frac{1}{p_i(x|y)} \\
&= -\sum_{i=1}^{n}\alpha_i \sum_{(x,y)\in W_i} p_i(x,y)\log\frac{p(x|y)}{p_i(x|y)}.
\end{aligned}
\tag{4.2}
$$

The second equality in (4.2) is valid since $p_i(x,y)\log(1/p(x|y)) = 0$, for $(x,y) \in R\setminus W_i$. We have $\sum_{(x,y)\in W_i} p_i(x,y) = 1$ for $i = 1,\cdots,n$. We can apply Lemma A (i) with $p_k = p_i(x,y)$ and $\xi_k = p(x|y)/p_i(x|y)$, $((x,y) \in W_i)$ to obtain

$$
\begin{aligned}
0 &\le \log\left[\sum_{(x,y)\in W_i} p_i(x,y)\frac{p(x|y)}{p_i(x|y)}\right] - \sum_{(x,y)\in W_i} p_i(x,y)\log\frac{p(x|y)}{p_i(x|y)} \\
&= \log T_i - \sum_{(x,y)\in W_i} p_i(x,y)\log\frac{p(x|y)}{p_i(x|y)} \\
&\le \log\left[\sum_{(u,v)\in W_i} p_i(u,v)\frac{p(u|v)}{p_i(u|v)}\sum_{(x,y)\in W_i} p_i(x,y)\frac{p_i(x|y)}{p(x|y)}\right] \\
&= \log T_i + \log\left[\sum_{(x,y)\in W_i}\frac{p_i(y)p_i^2(x|y)}{p(x|y)}\right], \quad (i = 1,\cdots,n).
\end{aligned}
$$

On multiplying this by $\alpha_i > 0$ and summing over $i = 1,\cdots,n$, we get by (4.2) that

$$
\begin{aligned}
0 &\le \log T + H(X|Y) - \sum_{i=1}^{n}\alpha_i H(X_i|Y_i) \\
&\le \log T + \sum_{i=1}^{n}\alpha_i \log\left[\sum_{(x,y)\in W_i}\frac{p_i(y)p_i^2(x|y)}{p(x|y)}\right].
\end{aligned}
$$

This is equivalent to the first two inequalities in (4.1). The third and fourth follow by (1.3) and (1.4), respectively.

Remark. Note that $0 < T_i \le 1$ for all $i = 1, \cdots, n$. It can happen that $T_i < 1$ for some indices i. Thus we have $0 < T \le 1$ and $-\log T \ge 0$. So the first inequality in (4.1) implies

$$H(X|Y) \ge \sum_{i=1}^{n} \alpha_i H(X_i|Y_i).$$

Theorem 4.2. *Let (X_i, Y_i) $(1 \le i \le n)$ and (X, Y) be as in Theorem 3.1. Define*

$$M_i := \max_{(x,y)\in W_i} \{p_i(x|y)/p(x|y)\}\ ,\ m_i := \min_{(x,y)\in W_i} \{p_i(x|y)/p(x|y)\} \quad (i = 1, \cdots, n).$$

Then

$$\begin{aligned} 0 &\le \log T + H(X|Y) - \sum_{i=1}^{n} \alpha_i H(X_i|Y_i) \\ &\le \sum_{i=1}^{n} \alpha_i \log\left[\frac{1}{4}(M_i - m_i)^2 T_i^2 + 1\right] \\ &= \log\left\{\prod_{i=1}^{n}\left[\frac{1}{4}(M_i - m_i)^2 T_i^2 + 1\right]^{\alpha_i}\right\} \\ &\le \frac{1}{\ln b}\left\{\prod_{i=1}^{n}\left[\frac{1}{4}(M_i - m_i)^2 T_i^2 + 1\right]^{\alpha_i} - 1\right\} \\ &\le \frac{1}{4\ln b}\sum_{i=1}^{n} \alpha_i (M_i - m_i)^2 T_i^2. \end{aligned} \tag{4.3}$$

Also we have

$$\begin{aligned} 0 &\le \log T + H(X|Y) - \sum_{i=1}^{n} \alpha_i H(X_i|Y_i) \le \sum_{i=1}^{n} \alpha_i \log\left[\frac{1}{4}\left(\sqrt{\rho_i} + \frac{1}{\sqrt{\rho_i}}\right)^2\right] \\ &= \log\left[\frac{1}{4}\prod_{i=1}^{n}\left(\sqrt{\rho_i} + \frac{1}{\sqrt{\rho_i}}\right)^{2\alpha_i}\right] \le \frac{1}{\ln b}\left[\frac{1}{4}\prod_{i=1}^{n}\left(\sqrt{\rho_i} + \frac{1}{\sqrt{\rho_i}}\right)^{2\alpha_i} - 1\right] \\ &\le \frac{1}{4\ln b}\sum_{i=1}^{n} \alpha_i \left(\sqrt{\rho_i} - \frac{1}{\sqrt{\rho_i}}\right)^2, \end{aligned} \tag{4.4}$$

where $\rho_i = M_i/m_i$ $(i = 1, \cdots, n)$.

Proof. To prove (4.3), note that for a fixed index i we have

$$\begin{aligned} &T_i \sum_{(x,y)\in W_i} \frac{p_i(y)p_i^2(x|y)}{p(x|y)} - 1 \\ &= \sum_{(x,y)\in W_i} p_i(y)p(x|y) \sum_{(u,v)\in W_i} p_i(v)p(u|v)\left(\frac{p_i(u|v)}{p(u|v)}\right)^2 \\ &\quad - \left(\sum_{(x,y)\in W_i} p_i(y)p(x|y)\frac{p_i(x|y)}{p(x|y)}\right)^2. \end{aligned}$$

Now apply Lemma B with $p_k = p_i(y)p(x|y)$ and $a_k = b_k = p_i(x|y)/p(x|y)$, $((x,y) \in W_i)$. We obtain

$$T_i \sum_{(x,y)\in W_i} \frac{p_i(y)p_i^2(x|y)}{p(x|y)} - 1 \leq \frac{1}{4}(M_i - m_i)^2 T_i^2, \quad (i = 1, \cdots, n),$$

or equivalently

$$\log T_i + \log\left[\sum_{(x,y)\in W_i} \frac{p_i(y)p_i^2(x|y)}{p(x|y)}\right] \leq \log\left[\frac{1}{4}(M_i - m_i)^2 T_i^2 + 1\right], \quad (i = 1, \cdots, n).$$

Multiply this by $\alpha_i > 0$, sum over $i = 1, \cdots, n$, then combine with (4.1). This gives the first two inequalities in (4.3). The rest of (4.3) follows by (1.3) and (1.4). Furthermore, by Lemma A (ii) with $p_k = p_i(x,y)$ and $\xi_k = p(x|y)/p_i(x|y)$, $((x,y) \in W_i)$ we have

$$\rho_i = \max_{j,k}\{\xi_j/\xi_k\} = (1/m_i)/(1/M_i) = M_i/m_i, \quad (i = 1, \cdots, n)$$

and

$$\begin{aligned} 0 &\leq \log\left[\sum_{(x,y)\in W_i} p_i(x,y)\frac{p(x|y)}{p_i(x|y)}\right] - \sum_{(x,y)\in W_i} p_i(x,y)\log\frac{p(x|y)}{p_i(x|y)} \\ &= \log T_i - \sum_{(x,y)\in W_i} p_i(x,y)\log\frac{p(x|y)}{p_i(x|y)} \\ &\leq \log\left[\frac{1}{4}\left(\sqrt{\rho_i} + \frac{1}{\sqrt{\rho_i}}\right)^2\right], \quad (i = 1, \cdots, n). \end{aligned}$$

Multiply this by $\alpha_i > 0$ and sum over $i = 1, \cdots, n$. By (4.2), we get the first two inequalities in (4.4). The rest of (4.4) follows by (1.3) and (1.4).

Remark. Let $\max_{1\leq i\leq n}\{M_i - m_i\} = M - m$. With an argument similat to that in the discussion following Theorem 2.2, we get

$$\begin{aligned} 0 &\leq \log T + H(X|Y) - \sum_{i=1}^n \alpha_i H(X_i|Y_i) \leq \log\left\{\prod_{i=1}^n \left[\frac{1}{4}(M-m)^2 T_i^2 + 1\right]^{\alpha_i}\right\} \\ &\leq \frac{(M-m)^2}{4\ln b}\sum_{i=1}^n \alpha_i S_i^2 \end{aligned}$$

and also

$$\begin{aligned} 0 &\leq \log T + H(X|Y) - \sum_{i=1}^n \alpha_i H(X_i|Y_i) \leq \log\left[\frac{1}{4}\left(\sqrt{\rho} + \frac{1}{\sqrt{\rho}}\right)^2\right] \\ &\leq \frac{1}{4\ln b}\left(\sqrt{\rho} - \frac{1}{\sqrt{\rho}}\right)^2, \end{aligned}$$

where $\rho = \max_{1\leq i\leq n}\{\rho_i\}$. Furthermore, if $p_i(x,y) > 0$ for all indices i and all $(x,y) \in R$, then $T_i = 1$ for all i and we have

$$0 \leq H(X|Y) - \sum_{i=1}^n \alpha_i H(X_i|Y_i) \leq \log\left[\frac{1}{4}(M-m)^2 + 1\right] \leq \frac{1}{4\ln b}(M-m)^2.$$

In that case, if

$$\max_{(x,y),(u,v)\in R}\left\{\max_{1\le i\le n}\left\{\left|\frac{p_i(x|y)}{p(x|y)}-\frac{p_i(u|v)}{p(u|v)}\right|\right\}\right\}\le 2\sqrt{b^{\varepsilon}-1}$$

for some $\varepsilon > 0$, then

$$0 \le H(X|Y) - \sum_{i=1}^{n} \alpha_i H(X_i|Y_i) \le \varepsilon.$$

In the case $n = 2$ this strengthens [3, Cor. 4.6], where the same conclusion is given but with $\sqrt{2\varepsilon \ln b}$ in place of $2\sqrt{b^{\varepsilon}-1}$.

As with entropy, the upper bounds on $H(X|Y) - \sum_{i=1}^{n} \alpha_i H(X_i|Y_i)$ derived in Theorems 4.1 and 4.2 depend on the values α_i. We have the following result which does not.

Theorem 4.3. *Under the conditions of Theorem 3.1,*

$$\begin{aligned}-\log T \le H(X|Y) &- \sum_{i=1}^{n} \alpha_i H(X_i|Y_i)\\ &\le \frac{1}{\ln b}\left[\sum_{(x,y)\in R}\max_{1\le i\le n}\{p_i(y)\}\max_{1\le i\le n}\{p_i(x|y)\}-1\right].\end{aligned}$$

Proof. We have

$$\begin{aligned}\sum_{i=1}^{n}\alpha_i p_i(y)p_i^2(x|y)/p(x|y) &= p(y)\sum_{i=1}^{n}\alpha_i p_i(y)p_i^2(x|y)\Big/\left(\sum_{i=1}^{n}\alpha_i p_i(x,y)\right)\\ &= p(y)\times\frac{\sum_{i=1}^{n}\alpha_i p_i(y)p_i^2(x|y)}{\sum_{i=1}^{n}\alpha_i p_i(y)p_i(x|y)}.\end{aligned}\tag{4.5}$$

By Lemma 2.4 with $t_i = p_i(y)$, $u_i = p_i(x|y)$, the second factor on the right–hand side in the second line of (4.5) is less than or equal to $\max_{1\le i\le n}\{p_i(x|y)\}$. Also

$$p(y) = \sum_{i=1}^{n}\alpha_i p_i(y) \le \max_{1\le i\le n}\{p_i(y)\}.$$

The desired result follows from (4.1). □

5. The Mutual Information for a Mixed Population

The mutual information of two discrete random variables X, Y is defined by

$$I(X,Y) := H(X) - H(X|Y) = \sum_{x,y} p(x,y)\log\frac{p(x,y)}{p(x)p(y)} = H(Y) - H(Y|X).$$

The summation in the formula above is over those pairs (x, y) for which $p(x, y) > 0$. Suppose that (X_1, Y_1), (X_1, Y_1) and (X, Y) are pairs of random variables such that

$$p(x) = \alpha p_1(x) + (1-\alpha)p_2(x), \quad (0 < \alpha < 1)$$

and that

$$p_1(y|x) = p_2(y|x) = p(y|x).$$

It is well–known that in this case we have

$$\alpha I(X_1, Y_1) + (1-\alpha) I(X_2, Y_2) \leq I(X, Y),$$

that is, for fixed transition probabilities, the mutual information $I(X, Y)$ is a concave functional of the input probabilities (see McEliece [8, Theorem 1.6]).

By contrast, if the roles of the input and transition probabilities are reversed, that is, if

$$p_1(x) = p_2(x) = p(x)$$

and

$$p(y|x) = \alpha p_1(y|x) + (1-\alpha) p_2(y|x), \quad (0 < \alpha < 1),$$

then

$$\alpha I(X_1, Y_1) + (1-\alpha) I(X_2, Y_2) \geq I(X, Y).$$

That is, for fixed input probabilities and $p(y|x)$ a convex linear combination of $p_1(y|x)$ and $p_2(y|x)$, we have that $I(X, Y)$ is a convex functional of the transition probabilities ([8, Theorem 1.7]). It should be noted that these facts are proved in [8] under the tacit assumption that the probabilities involved are strictly positive for all x and y.

In this section we shall see that the n–pair generalizations of the above two results follow from our results on entropy and conditional entropy.

We consider n pairs (X_i, Y_i) $(i = 1, \cdots, n)$ of discrete random variables with common finite range, and a new pair (X, Y) which we obtain as a convex linear combination of those n pairs. The common range of X_i $(i = 1, \cdots, n)$ and X is denoted by U, and that of Y_i $(i = 1, \cdots, n)$ and Y by V.

Theorem 5.1. *Let (X_i, Y_i) $(i = 1, \cdots, n)$, (X, Y), U and V be as above. Define sets U_i and V_i by*

$$U_i := \{x \in U : p_i(x) > 0\}, \; V_i := \{y \in V : p_i(y) > 0\}, \quad i = 1, \cdots, n.$$

Suppose that for given $\alpha_i > 0$ $(i = 1, \cdots, n)$ with $\sum_{i=1}^{n} \alpha_i = 1$, the conditions

$$p(x) = \sum_{i=1}^{n} \alpha_i p_i(x) > 0, \; \forall x \in U \tag{5.1}$$

and

$$p_i(y|x) = p(y|x), \; \forall i = 1, \cdots, n, \; \forall y \in V \quad \text{whenever } x \in U_i \tag{5.2}$$

are satisfied. Then we have

$$\begin{aligned} -\log S &\leq I(X, Y) - \sum_{i=1}^{n} \alpha_i I(X_i, Y_i) \leq \sum_{i=1}^{n} \alpha_i \log\left[\sum_{y \in V_i} \frac{p_i^2(y)}{p(y)}\right] \\ &= \log\left[\prod_{i=1}^{n} \left(\sum_{y \in V_i} \frac{p_i^2(y)}{p(y)}\right)^{\alpha_i}\right] \leq \frac{1}{\ln b}\left[\prod_{i=1}^{n} \left(\sum_{y \in V_i} \frac{p_i^2(y)}{p(y)}\right)^{\alpha_i} - 1\right] \\ &\leq \frac{1}{\ln b}\left[\sum_{y \in V} \sum_{i=1}^{n} \alpha_i \frac{p_i^2(y)}{p(y)} - 1\right], \end{aligned}$$

where

$$S_i = \sum\nolimits_{y\in V_i} p(y),\ i=1,\cdots,n \ \text{ and } \ S=\prod_{i=1}^{n} S_i^{\alpha_i}.$$

Proof. We have

$$\begin{aligned} &I(X,Y)-\sum_{i=1}^{n}\alpha_i I(X_i,Y_i) \\ &= H(Y)-\sum_{i=1}^{n}\alpha_i H(Y_i) - \left[H(Y|X)-\sum_{i=1}^{n}\alpha_i H(Y_i|X_i)\right]. \end{aligned} \tag{5.3}$$

Using (5.1) and (5.2), we get for all $x\in U$ and $y\in V$ that

$$p(x,y)=p(x)p(y|x)=\left[\sum_{i=1}^{n}\alpha_i p_i(x)\right]p(y|x)=\sum_{i=1}^{n}\alpha_i p_i(x,y),$$

since $p_i(x)p(y|x)=p_i(x)p_i(y|x)=p_i(x,y)$ when $x\in U_i$, and $p_i(x)p(y|x)=0=p_i(x,y)$ when $x\notin U_i$. Also, for all $y\in V$ we have

$$\begin{aligned} p(y) &= \sum_{x\in U} p(x,y) = \sum_{x\in U}\left(\sum_{i=1}^{n}\alpha_i p_i(x,y)\right) \\ &= \sum_{i=1}^{n}\alpha_i\left(\sum_{x\in U} p_i(x,y)\right) = \sum_{i=1}^{n}\alpha_i p_i(y). \end{aligned}$$

Further, put

$$W=\{(x,y): p(x,y)>0 \quad \text{and} \quad W_i=\{(x,y): p_i(x,y)>0 \quad (i=1,\cdots,n).$$

We have

$$\begin{aligned} H(Y|X) &= \sum_{(x,y)\in W} p(x,y)\log\frac{1}{p(y|x)} = \sum_{(x,y)\in W}\left(\sum_{i=1}^{n}\alpha_i p_i(x,y)\right)\log\frac{1}{p(y|x)} \\ &= \sum_{i=1}^{n}\alpha_i\left(\sum_{(x,y)\in W_i} p_i(x,y)\log\frac{1}{p_i(y|x)}\right) = \sum_{i=1}^{n}\alpha_i H(Y_i|X_i). \end{aligned}$$

Therefore the difference in brackets in (5.3) is equal to zero and we have

$$I(X,Y)-\sum_{i=1}^{n}\alpha_i I(X_i,Y_i) = H(Y)-\sum_{i=1}^{n}\alpha_i H(Y_i). \tag{5.4}$$

Since (5.4) holds, we can apply Theorem 2.1 with Y_i $(i=1,\cdots,n)$ and Y in place of X_i $(i=1,\cdots,n)$ and X (respectively) to obtain the desired result.

Theorem 5.2. *Under the assumptions and with the notation of the previous theorem, we have*

$$0 \leq \log S + I(X,Y) - \sum_{i=1}^{n} \alpha_i I(X_i, Y_i) \leq \sum_{i=1}^{n} \alpha_i \log\left[\frac{1}{4}(M_i - m_i)^2 S_i^2 + 1\right]$$
$$= \log\left\{\prod_{i=1}^{n}\left[\frac{1}{4}(M_i - m_i)^2 S_i^2 + 1\right]^{\alpha_i}\right\} \leq \frac{1}{\ln b}\left\{\prod_{i=1}^{n}\left[\frac{1}{4}(M_i - m_i)^2 S_i^2 + 1\right]^{\alpha_i} - 1\right\}$$
$$\leq \frac{1}{4\ln b}\sum_{i=1}^{n} \alpha_i (M_i - m_i)^2 S_i^2,$$

where

$$M_i := \max_{y\in V_i}\{p_i(y)/p(y)\}, \ m_i := \min_{y \in V_i}\{p_i(y)/p(y)\} \quad (i = 1, \cdots, n).$$

Also,

$$0 \leq \log S + (X,Y) - \sum_{i=1}^{n} \alpha_i I(X_i, Y_i) \leq \sum_{i=1}^{n} \alpha_i \log\left[\frac{1}{4}\left(\sqrt{\rho_i} + \frac{1}{\sqrt{\rho_i}}\right)^2\right]$$
$$= \log\left[\frac{1}{4}\prod_{i=1}^{n}\left(\sqrt{\rho_i} + \frac{1}{\sqrt{\rho_i}}\right)^{2\alpha_i}\right] \leq \frac{1}{\ln b}\left[\frac{1}{4}\prod_{i=1}^{n}\left(\sqrt{\rho_i} + \frac{1}{\sqrt{\rho_i}}\right)^{2\alpha_i} - 1\right]$$
$$\leq \frac{1}{4\ln b}\sum_{i=1}^{n}\alpha_i\left(\sqrt{\rho_i} - \frac{1}{\sqrt{\rho_i}}\right)^2,$$

where $\rho_i = M_i/m_i$ $(i = 1, \cdots, n)$.

Proof. Use (5.4) and apply Theorem 2.2 with Y_i $(i = 1, \cdots, n)$ and Y in place of X_i $(i = 1, \cdots, n)$ and X (respectively).

Remark. As in Section 2, we have $\max_{1 \leq i \leq n}\{M_i - m_i\} = M - m$ and

$$0 \leq \log S + I(X,Y) - \sum_{i=1}^{n} \alpha_i I(X_i, Y_i) \leq \log\left\{\prod_{i=1}^{n}\left[\frac{1}{4}(M - m)^2 S_i^2 + 1\right]^{\alpha_i}\right\}$$
$$\leq \frac{(M-m)^2}{4\ln b}\sum_{i=1}^{n}\alpha_i S_i^2.$$

Also

$$0 \leq \log S + I(X,Y) - \sum_{i=1}^{n} \alpha_i I(X_i, Y_i) \leq \log\left[\frac{1}{4}\left(\sqrt{\rho} + \frac{1}{\sqrt{\rho}}\right)^2\right]$$
$$\leq \frac{1}{4\ln b}\left(\sqrt{\rho} - \frac{1}{\sqrt{\rho}}\right)^2,$$

where $\rho = \max_{1 \leq i \leq n}\{\rho_i\}$. Moreover, suppose $p_i(y) > 0$ for all $i = 1, \cdots, n$ and all $y \in V$. If $M - m \leq 2\sqrt{b^\varepsilon - 1}$ for some $\varepsilon > 0$, then

$$0 \leq I(X,Y) - \sum_{i=1}^{n} \alpha_i I(X_i, Y_i) \leq \varepsilon.$$

In the case $n = 2$ this result subsumes [3, Cor. 5.8], which gives the result under the more demanding condition $M - m \leq \sqrt{2\varepsilon \ln b}$.

Theorem 5.3. *Under the assumptions of Theorem 5.1, we have*

$$-\log S \le I(X,Y) - \sum_{i=1}^{n} \alpha_i I(X_i, Y_i) \le \frac{1}{\ln b}\left[\sum_{y\in V} \max_{1\le i\le n}\{p_i(y)\} - 1\right].$$

Proof. The result follows from Theorem 2.5.

Theorem 5.4. *Let (X_i, Y_i) $(1 \le i \le n)$ and (X, Y) be pairs of discrete random variables with common finite range R. Suppose that for given $\alpha_i > 0$ $(i = 1, \cdots, n)$ with $\sum_{i=1}^{n} \alpha_i = 1$ the conditions*

$$p_i(x) = p(x) > 0, \quad \forall i = 1, \cdots, n, \ \forall x \in U, \tag{5.5}$$

$$p(y|x) = \sum_{i=1}^{n} \alpha_i p_i(y|x), \quad \forall (x,y) \in R \tag{5.6}$$

are satisfied. Then

$$\begin{aligned} -\log T &\le \sum_{i=1}^{n} \alpha_i I(X_i|Y_i) - I(X,Y) \le \sum_{i=1}^{n} \alpha_i \log\left[\sum_{(x,y)\in W_i} \frac{p_i(y)p_i^2(x|y)}{p(x|y)}\right] \\ &= \log\left[\prod_{i=1}^{n}\left(\sum_{(x,y)\in W_i} \frac{p_i(y)p_i^2(x|y)}{p(x|y)}\right)^{\alpha_i}\right] \\ &\le \frac{1}{\ln b}\left[\prod_{i=1}^{n}\left(\sum_{(x,y)\in W_i} \frac{p_i(y)p_i^2(x|y)}{p(x|y)}\right)^{\alpha_i} - 1\right] \\ &\le \frac{1}{\ln b}\left[\sum_{(x,y)\in R}\sum_{i=1}^{n} \alpha_i \frac{p_i(y)p_i^2(x|y)}{p(x|y)} - 1\right], \end{aligned}$$

where

$$W_i = \{(x,y) : p_i(x,y) > 0\}, \quad T_i = \sum_{(x,y)\in W_i} p_i(y)p(x|y) \quad i = 1, \cdots, n$$

and $T = \prod_{i=1}^{n} T_i^{\alpha_i}$.

Proof. Since (5.5) and (5.6) hold for all $(x, y) \in R$ we have

$$\begin{aligned} p(x,y) = p(x)p(y|x) &= p(x)\left[\sum_{i=1}^{n} \alpha_i p_i(y|x)\right] = \sum_{i=1}^{n} \alpha_i p(x) p_i(y|x) \\ &= \sum_{i=1}^{n} \alpha_i p_i(x) p(y|x) = \sum_{i=1}^{n} \alpha_i p_i(x,y). \end{aligned}$$

Also it follows from (5.5) that $H(X_i) = H(X)$ for all $i = 1, \cdots, n$, so that

$$\begin{aligned} \sum_{i=1}^{n} \alpha_i I(X_i, Y_i) - I(X,Y) &= \sum_{i=1}^{n} \alpha_i [H(X_i) - H(X_i|Y_i)] - [H(X) - H(X|Y)] \\ &= H(X|Y) - \sum_{i=1}^{n} \alpha_i H(X_i|Y_i). \end{aligned} \tag{5.7}$$

The result follows by Theorem 4.1.

Remark. It is clear from (5.7) that Theorems 4.2 and 4.3 apply when we substitute $\sum_{i=1}^{n} \alpha_i I(X_i, Y_i) - I(X,Y)$ in place of $H(X|Y) - \sum_{i=1}^{n} \alpha_i H(X_i|Y_i)$.

References

1. S. S. Dragomir and C. J. Goh, *A counterpart of Jensen's discrete inequality for differentiable convex mappings and applications in information theory*, Math. Comput. Modelling **24 (2)** (1996), 1–11.
2. S. S. Dragomir and C. J. Goh, *Some bounds on entropy measures in information theory*, Appl. Math. Lett. **10 (3)** (1997), 23–28.
3. S. S. Dragomir and C. J. Goh, *Further counterparts of some inequalities in information theory*, Rend. Sem. Mat. Univ. Pol. Torino **56** (1998), 35–50.
4. M. Matić, C. E. M. Pearce and J. Pečarić, *Improvements of some bounds on entropy measures in information theory*, Math. Inequal. and Appl. **1** (1998), 295–304.
5. M. Matić, C. E. M. Pearce and J. Pečarić, *Further improvements of some bounds on entropy measures in information theory*, Math. Ineq. and Applic. **2** (1999), 599–611.
6. M. Matić, C. E. M. Pearce and J. Pečarić, *Bounds on entropy measures for mixed populations*, in Analytic & Geometric Inequalities & Applications, Eds Th. Rassias and H. M. Srivastava, Kluwer (1999), 231–244.
7. M. Matić, C. E. M. Pearce and J. Pečarić, *Refinements of some bounds in information theory*, ANZIAM J. **42** (2001), 387–398.
8. R. J. McEliece, *The Theory of Information and Coding*, Addison-Wesley, 1977.
9. D. S. Mitrinović, J. E. Pečarić and A. M. Fink, *Classical and New Inequalities in Analysis*, Kluwer Acad. Publ., 1993.

In: Inequality Theory and Applications, Volume 4
Editors: Yeol Je Cho et al., pp.111–118

ISBN 1-59454-874-9

ALGEBRA HOMOMORPHISMS IN C^*-ALGEBRAS

CHUN-GIL PARK[1]

ABSTRACT. It is shown that every almost C^*-algebra homomorphism $h : \mathcal{A} \to \mathcal{B}$ of a unital C^*-algebra $\mathcal{A}$ to a unital C^*-algebra $\mathcal{B}$ is a C^*-algebra homomorphism when $h(rx) = rh(x)$ $(r > 1)$ for all $x \in \mathcal{A}$, and that every almost linear mapping $h : \mathcal{A} \to \mathcal{B}$ is a C^*-algebra homomorphism when $h(2^n uy) = h(2^n u)h(y)$, $h(3^n uy) = h(3^n u)h(y)$, or $h(q^n uy) = h(q^n u)h(y)$ for all unitaries $u \in \mathcal{A}$, all $y \in \mathcal{A}$, and $n = 0, 1, \cdots$. Here the numbers $2, 3, q$ depend on the functional equations given in the almost linear mappings.

1. Introduction

Let X and Y be Banach spaces with norms $||\cdot||$ and $\|\cdot\|$, respectively. Consider $f : X \to Y$ to be a mapping such that $f(tx)$ is continuous in $t \in \mathbb{R}$ for each fixed $x \in X$. Assume that there exist constants $\theta \geq 0$ and $p \in [0, 1)$ such that

$$\|f(x+y) - f(x) - f(y)\| \leq \theta(||x||^p + ||y||^p)$$

for all $x, y \in X$. Rassias [8] showed that there exists a unique $\mathbb{R}$-linear mapping $T : X \to Y$ such that

$$\|f(x) - T(x)\| \leq \frac{2\theta}{2 - 2^p}||x||^p$$

for all $x \in X$. Găvruta [1] generalized the Rassias' result: Let G be an abelian group and Y a Banach space. Denote by $\varphi : G \times G \to [0, \infty)$ a function such that

$$\widetilde{\varphi}(x, y) = \sum_{j=0}^{\infty} 2^{-j} \varphi(2^j x, 2^j y) < \infty$$

for all $x, y \in G$. Suppose that $f : G \to Y$ is a mapping satisfying

$$\|f(x+y) - f(x) - f(y)\| \leq \varphi(x, y)$$

for all $x, y \in G$. Then there exists a unique additive mapping $T : G \to Y$ such that

$$\|f(x) - T(x)\| \leq \frac{1}{2}\widetilde{\varphi}(x, x)$$

Received October 31, 2003.
2000*Mathematics Subject Classification*. Primary 47C15, 46L05, 39B52.
Key words and phrases. C^*-algebra homomorphism, stability, linear functional equation.
[1] Department of Mathematics, Chungnam National University, Daejeon 305-764, Korea.
E-mail: cgpark@math.cnu.ac.kr

for all $x \in G$. Park [6] applied the Găvruta's result to linear functional equations in Banach modules over a C^*-algebra.

Jun and Lee [3] proved the stability of Jensen's equation. Recently, Trif [9] proved the stability of a functional equation deriving from an inequality of Popoviciu for convex functions.

Johnson [2, Theorem 7.2] also investigated almost algebra $*$-homomorphisms between Banach $*$-algebras : Suppose that $\mathcal{U}$ and $\mathcal{B}$ are Banach $*$-algebras which satisfy the conditions of [2, Theorem 3.1]. Then for each positive ϵ and K there is a positive δ such that if $T \in L(\mathcal{U}, \mathcal{B})$ with $\|T\| < K$, $\|T^\vee\| < \delta$ and $\|T(x^*)^* - T(x)\| \le \delta\|x\| \quad (x \in \mathcal{U})$ then there is a $*$-homomorphism $T' : \mathcal{U} \to \mathcal{B}$ with $\|T - T'\| < \epsilon$. Here $L(\mathcal{U}, \mathcal{B})$ is the space of bounded linear maps from $\mathcal{U}$ into $\mathcal{B}$, and $T^\vee(x, y) = T(xy) - T(x)T(y) \quad (x, y \in \mathcal{U})$. See [2] for details.

Throughout this paper, let $\mathcal{A}$ be a unital C^*-algebra with norm $||\cdot||$ and unit e, and $\mathcal{B}$ a unital C^*-algebra with norm $\|\cdot\|$. Let $\mathcal{U}(\mathcal{A}) = \{u \in \mathcal{A} \mid uu^* = u^*u = e\}$. Let $q = \frac{l(d-1)}{d-l}$ for positive integers l, d with $2 \le l \le d-1$, and r a real number greater than 1.

Using the stability methods, we prove that every almost C^*-algebra homomorphism $h : \mathcal{A} \to \mathcal{B}$ is a C^*-algebra homomorphism when $h(rx) = rh(x)$ for all $x \in \mathcal{A}$, and that every almost linear mapping $h : \mathcal{A} \to \mathcal{B}$ is a C^*-algebra homomorphism when $h(2^n uy) = h(2^n u)h(y)$, $h(3^n uy) = h(3^n u)h(y)$, or $h(q^n uy) = h(q^n u)h(y)$ for all $u \in \mathcal{U}(\mathcal{A})$, all $y \in \mathcal{A}$ and $n = 0, 1, \cdots$.

2. C^*-Algebra Homomorphisms between Unital C^*-Algebras

We are going to investigate C^*-algebra homomorphisms between unital C^*-algebras associated with the Cauchy functional equation.

Theorem 2.1. *Let $h : \mathcal{A} \to \mathcal{B}$ be a mapping satisfying $h(rx) = rh(x)$ for all $x \in \mathcal{A}$ for which there exists a function $\varphi : \mathcal{A}^4 \to [0, \infty)$ such that*

(i) $\widetilde{\varphi}(x, y, z, w) := \sum_{j=0}^{\infty} r^{-j}\varphi(r^j x, r^j y, r^j z, r^j w) < \infty$,

(ii) $\|h(\mu x + \mu y + zw) - \mu h(x) - \mu h(y) - h(z)h(w)\| \le \varphi(x, y, z, w)$,

(iii) $\|h(r^n u^*) - h(r^n u)^*\| \le \varphi(r^n u, r^n u, 0, 0)$

for all $\mu \in \mathbb{T}^1 := \{\lambda \in \mathbb{C} \mid |\lambda| = 1\}$, $u \in \mathcal{U}(\mathcal{A})$, $n = 0, 1, \cdots$, and $x, y, z, w \in \mathcal{A}$.

Then the mapping $h : \mathcal{A} \to \mathcal{B}$ is a C^*-algebra homomorphism.

Proof. Since $h(0) = rh(0)$, $h(0) = 0$. Put $z = w = 0$ and $\mu = 1 \in \mathbb{T}^1$ in (ii). By (ii) and the assumption that $h(rx) = rh(x)$ for all $x \in \mathcal{A}$,

$$\begin{aligned}\|h(x+y) - h(x) - h(y)\| &= \frac{1}{r^n}\|h(r^n x + r^n y) - h(r^n x) - h(r^n y)\| \\ &\le \frac{1}{r^n}\varphi(r^n x, r^n y, 0, 0),\end{aligned}$$

which tends to zero as $n \to \infty$ by (i). So

$$h(x+y) = h(x) + h(y)$$

for all $x, y \in \mathcal{A}$. Put $y = z = w = 0$ in (ii). By (ii) and the assumption that $h(rx) = rh(x)$ for all $x \in \mathcal{A}$,

$$\|h(\mu x) - \mu h(x)\| = \frac{1}{r^n}\|h(r^n \mu x) - \mu h(r^n x)\| \le \frac{1}{r^n}\varphi(r^n x, 0, 0, 0),$$

which tends to zero as $n \to \infty$ by (i). So

$$h(\mu x) = \mu h(x) \tag{2.1}$$

for all $\mu \in \mathbb{T}^1$ and $x \in \mathcal{A}$.

Now, let $\lambda \in \mathbb{C}$ ($\lambda \neq 0$) and M an integer greater than $4|\lambda|$. Then $|\frac{\lambda}{M}| < \frac{1}{4} < 1 - \frac{2}{3} = \frac{1}{3}$. By [4, Theorem 1], there exist three elements $\mu_1, \mu_2, \mu_3 \in \mathbb{T}^1$ such that $3\frac{\lambda}{M} = \mu_1 + \mu_2 + \mu_3$. And $h(x) = h(3 \cdot \frac{1}{3}x) = 3h(\frac{1}{3}x)$ for all $x \in \mathcal{A}$. So $h(\frac{1}{3}x) = \frac{1}{3}h(x)$ for all $x \in \mathcal{A}$. Thus, by (2.1)

$$\begin{aligned} h(\lambda x) &= h\Big(\frac{M}{3} \cdot 3\frac{\lambda}{M}x\Big) = M \cdot h\Big(\frac{1}{3} \cdot 3\frac{\lambda}{M}x\Big) = \frac{M}{3}h\Big(3\frac{\lambda}{M}x\Big) \\ &= \frac{M}{3}h(\mu_1 x + \mu_2 x + \mu_3 x) = \frac{M}{3}(h(\mu_1 x) + h(\mu_2 x) + h(\mu_3 x)) \\ &= \frac{M}{3}(\mu_1 + \mu_2 + \mu_3)h(x) = \frac{M}{3} \cdot 3\frac{\lambda}{M}h(x) \\ &= \lambda h(x) \end{aligned}$$

for all $x \in \mathcal{A}$. Hence

$$h(\zeta x + \eta y) = h(\zeta x) + h(\eta y) = \zeta h(x) + \eta h(y)$$

for all $\zeta, \eta \in \mathbb{C}(\zeta, \eta \neq 0)$ and all $x, y \in \mathcal{A}$. And $h(0x) = 0 = 0h(x)$ for all $x \in \mathcal{A}$. So the mapping $h : \mathcal{A} \to \mathcal{B}$ is a $\mathbb{C}$-linear mapping.

Put $x = y = 0$ in (ii). By (ii) and the assumption that $h(rx) = rh(x)$ for all $x \in \mathcal{A}$,

$$\begin{aligned} \|h(zw) - h(z)h(w)\| &= \frac{1}{r^{2n}}\|h(r^n z \cdot r^n w) - h(r^n z)h(r^n w)\| \\ &\leq \frac{1}{r^{2n}}\varphi(0, 0, r^n z, r^n w) \\ &\leq \frac{1}{r^n}\varphi(0, 0, r^n z, r^n w), \end{aligned}$$

which tends to zero as $n \to \infty$ by (i). So

$$h(zw) = h(z)h(w)$$

for all $z, w \in \mathcal{A}$. By (iii) and the assumption that $h(rx) = rh(x)$ for all $x \in \mathcal{A}$,

$$\|h(u^*) - h(u)^*\| = \frac{1}{r^n}\|h(r^n u^*) - h(r^n u)^*\| \leq \frac{1}{r^n}\varphi(r^n u, r^n u, 0, 0),$$

which tends to zero as $n \to \infty$ by (i). So

$$h(u^*) = h(u)^*$$

for all $u \in \mathcal{U}(\mathcal{A})$. Since $h : \mathcal{A} \to \mathcal{B}$ is $\mathbb{C}$-linear and each $x \in \mathcal{A}$ is a finite linear combination of unitary elements (see [5, Theorem 4.1.7]), i.e., $x = \sum_{j=1}^m \lambda_j u_j$ ($\lambda_j \in \mathbb{C}, u_j \in \mathcal{U}(\mathcal{A})$),

$$\begin{aligned} h(x^*) &= h\Big(\sum_{j=1}^m \overline{\lambda_j} u_j^*\Big) = \sum_{j=1}^m \overline{\lambda_j} h(u_j^*) = \sum_{j=1}^m \overline{\lambda_j} h(u_j)^* \\ &= \Big(\sum_{j=1}^m \lambda_j h(u_j)\Big)^* = \Big(\sum_{j=1}^m \lambda_j u_j\Big)^* = h(x)^* \end{aligned}$$

for all $x \in \mathcal{A}$. Therefore, the mapping $h : \mathcal{A} \to \mathcal{B}$ is a C^*-algebra homomorphism. This completes the proof.

Corollary 2.2. *Let $h : \mathcal{A} \to \mathcal{B}$ be a mapping satisfying $h(rx) = rh(x)$ for all $x \in \mathcal{A}$ for which there exist constants $\theta \geq 0$ and $p \in [0, 1)$ such that*

$$\begin{aligned} &\|h(\mu x + \mu y + zw) - \mu h(x) - \mu h(y) - h(z)h(w)\| \\ &\quad \leq \theta(||x||^p + ||y||^p + ||z||^p + ||w||^p), \\ &\|h(r^n u^*) - h(r^n u)^*\| \leq 2r^{np}\theta \end{aligned}$$

for all $\mu \in \mathbb{T}^1$, $u \in \mathcal{U}(\mathcal{A})$, $n = 0, 1, \cdots$, and $x, y, z, w \in \mathcal{A}$. Then the mapping $h : \mathcal{A} \to \mathcal{B}$ is a C^*-algebra homomorphism.

Proof. Define $\varphi(x, y, z, w) = \theta(||x||^p + ||y||^p + ||z||^p + ||w||^p)$, and apply Theorem 2.1.

Theorem 2.3. *Let $h : \mathcal{A} \to \mathcal{B}$ be a mapping satisfying $h(rx) = rh(x)$ for all $x \in \mathcal{A}$ for which there exists a function $\varphi : \mathcal{A}^4 \to [0, \infty)$ satisfying* (i) *and* (iii) *such that*

(iv) $\|h(\mu x + \mu y + zw) - \mu h(x) - \mu h(y) - h(z)h(w)\| \leq \varphi(x, y, z, w)$

for $\mu = 1, i$, and $x, y, z, w \in \mathcal{A}$. If $h(tx)$ is continuous in $t \in \mathbb{R}$ for each fixed $x \in \mathcal{A}$, then the mapping $h : \mathcal{A} \to \mathcal{B}$ is a C^*-algebra homomorphism.

Proof. Put $z = w = 0$ and $\mu = 1$ in (iv). By the same reasoning as in the proof of Theorem 2.1, the mapping $h : \mathcal{A} \to \mathcal{B}$ is additive. By the same reasoning as in the proof of [8, Theorem], the additive mapping $h : \mathcal{A} \to \mathcal{B}$ is $\mathbb{R}$-linear.

Put $z = w = 0$ and $\mu = i$ in (iv). By the same method as in the proof of Theorem 2.1, one can obtain that

$$h(ix) = ih(x)$$

for all $x \in \mathcal{A}$. For each element $\lambda \in \mathbb{C}$, $\lambda = s + it$, where $s, t \in \mathbb{R}$. So

$$h(\lambda x) = h(sx + itx) = sh(x) + th(ix) = sh(x) + ith(x) = (s + it)h(x) = \lambda h(x)$$

for all $\lambda \in \mathbb{C}$ and $x \in \mathcal{A}$. So

$$h(\zeta x + \eta y) = h(\zeta x) + h(\eta y) = \zeta h(x) + \eta h(y)$$

for all $\zeta, \eta \in \mathbb{C}$, and $x, y \in \mathcal{A}$. Hence the additive mapping $h : \mathcal{A} \to \mathcal{B}$ is $\mathbb{C}$-linear.

The rest of the proof is the same as in the proof of Theorem 2.1. This completes the proof.

Park [7] investigated C^*-algebra homomorphisms between C^*-algebras. Under a weak condition, we are going to prove the same result.

Theorem 2.4. *Let $h : \mathcal{A} \to \mathcal{B}$ be a mapping satisfying $h(0) = 0$ and $h(2^n uy) = h(2^n u)h(y)$ for all $u \in \mathcal{U}(\mathcal{A})$, all $y \in \mathcal{A}$, and $n = 0, 1, \cdots$, for which there exists a function $\varphi : \mathcal{A} \times \mathcal{A} \to [0, \infty)$ such that*

(v) $\widetilde{\varphi}(x, y) := \sum_{j=0}^{\infty} 2^{-j}\varphi(2^j x, 2^j y) < \infty$,

(vi) $\|h(\mu x + \mu y) - \mu h(x) - \mu h(y)\| \leq \varphi(x, y)$,

(vii) $\|h(2^n u^*) - h(2^n u)^*\| \leq \varphi(2^n u, 2^n u)$

for all $\mu \in \mathbb{T}^1$, $u \in \mathcal{U}(\mathcal{A})$, $n = 0, 1, \cdots$, and $x, y \in \mathcal{A}$. Assume $\lim_{n\to\infty} \frac{h(2^n e)}{2^n}$ is invertible. Then the mapping $h : \mathcal{A} \to \mathcal{B}$ is a C^*-algebra homomorphism.

Proof. Put $\mu = 1 \in \mathbb{T}^1$ in (vi). It follows from Găvruta Theorem [1] that there exists a unique additive mapping $H : \mathcal{A} \to \mathcal{B}$ such that

$$\|h(x) - H(x)| \leq \frac{1}{2}\widetilde{\varphi}(x, x)$$

for all $x \in \mathcal{A}$. The additive mapping $H : \mathcal{A} \to \mathcal{B}$ is given by

$$H(x) = \lim_{n\to\infty} \frac{1}{2^n} h(2^n x) \tag{2.2}$$

for all $x \in \mathcal{A}$. By the assumption, for each $\mu \in \mathbb{T}^1$,

$$\|h(2^n \mu x) - 2\mu h(2^{n-1}x)\| \le \varphi(2^{n-1}x, 2^{n-1}x)$$

for all $x \in \mathcal{A}$. And one can show that

$$\|\mu h(2^n x) - 2\mu h(2^{n-1}x)\| \le |\mu| \cdot \|h(2^n x) - 2h(2^{n-1}x)\| \le \varphi(2^{n-1}x, 2^{n-1}x)$$

for all $\mu \in \mathbb{T}^1$ and $x \in \mathcal{A}$. So

$$\begin{aligned}\|h(2^n \mu x) - \mu h(2^n x)\| \le & \|h(2^n \mu x) - 2\mu h(2^{n-1}x)\| + \|2\mu h(2^{n-1}x) - \mu h(2^n x)\| \\ \le & \varphi(2^{n-1}x, 2^{n-1}x) + \varphi(2^{n-1}x, 2^{n-1}x)\end{aligned}$$

for all $\mu \in \mathbb{T}^1$ and $x \in \mathcal{A}$. Thus $2^{-n}\|h(2^n\mu x) - \mu h(2^n x)\| \to 0$ as $n \to \infty$ for all $\mu \in \mathbb{T}^1$ and $x \in \mathcal{A}$. Hence

$$H(\mu x) = \lim_{n\to\infty} \frac{h(2^n \mu x)}{2^n} = \lim_{n\to\infty} \frac{\mu h(2^n x)}{2^n} = \mu H(x)$$

for all $\mu \in \mathbb{T}^1$ and $x \in \mathcal{A}$. By the same method as in the proof of Theorem 2.1, one can show that the unique additive mapping $H : \mathcal{A} \to \mathcal{B}$ is a $\mathbb{C}$-linear mapping.

By (2.2) and (vii), we get

$$H(u^*) = \lim_{n\to\infty} \frac{h(2^n u^*)}{2^n} = \lim_{n\to\infty} \frac{h(2^n u)^*}{2^n} = (\lim_{n\to\infty} \frac{h(2^n u)}{2^n})^* = H(u)^*$$

for all $u \in \mathcal{U}(\mathcal{A})$. By the same method as in the proof of Theorem 2.1, one can show that

$$H(x^*) = H(x)^*$$

for all $x \in \mathcal{A}$. Since $h(2^n uy) = h(2^n u)h(y)$ for all $u \in \mathcal{U}(\mathcal{A})$, $y \in \mathcal{A}$ and $n = 0, 1, \cdots$,

$$H(uy) = \lim_{n\to\infty} \frac{1}{2^n} h(2^n uy) = \lim_{n\to\infty} \frac{1}{2^n} h(2^n u)h(y) = H(u)h(y) \tag{2.3}$$

for all $u \in \mathcal{U}(\mathcal{A})$ and $y \in \mathcal{A}$. By the additivity of H and (3),

$$2^n H(uy) = H(2^n uy) = H(u(2^n y)) = H(u)h(2^n y)$$

for all $u \in \mathcal{U}(\mathcal{A})$ and $y \in \mathcal{A}$. Hence

$$H(uy) = \frac{1}{2^n} H(u)h(2^n y) = H(u)\frac{1}{2^n} h(2^n y) \tag{2.4}$$

for all $u \in \mathcal{U}(\mathcal{A})$ and $y \in \mathcal{A}$. Taking the limit in (2.4) as $n \to \infty$, we obtain

$$H(uy) = H(u)H(y) \tag{2.5}$$

for all $u \in \mathcal{U}(\mathcal{A})$ and $y \in \mathcal{A}$. Since H is $\mathbb{C}$-linear and each $x \in \mathcal{A}$ is a finite linear combination of unitary elements, i.e., $x = \sum_{j=1}^{m} \lambda_j u_j$ ($\lambda_j \in \mathbb{C}, u_j \in \mathcal{U}(\mathcal{A})$), it follows from (2.5) that

$$\begin{aligned} H(xy) &= H\left(\sum_{j=1}^{m} \lambda_j u_j y\right) = \sum_{j=1}^{m} \lambda_j H(u_j y) = \sum_{j=1}^{m} \lambda_j H(u_j) H(y) \\ &= H\left(\sum_{j=1}^{m} \lambda_j u_j\right) H(y) = H(x)H(y) \end{aligned}$$

for all $x, y \in \mathcal{A}$. By (2.3) and (2.5),

$$H(e)H(y) = H(ey) = H(e)h(y)$$

for all $y \in \mathcal{A}$. Since $\lim_{n\to\infty} \frac{h(2^n e)}{2^n} = H(e)$ is invertible,

$$H(y) = h(y)$$

for all $y \in \mathcal{A}$. Therefore, the mapping $h : \mathcal{A} \to \mathcal{B}$ is a C^*-algebra homomorphism. This completes the proof.

Now, we are going to investigate C^*-algebra homomorphisms between unital C^*-algebras associated with the Jensen functional equation.

Theorem 2.5. *Let $h : \mathcal{A} \to \mathcal{B}$ be a mapping satisfying $h(rx) = rh(x)$ for all $x \in \mathcal{A}$ for which there exists a function $\varphi : \mathcal{A}^4 \to [0, \infty)$ such that*

$$\begin{aligned} &\widetilde{\varphi}(x, y, z, w) := \sum_{j=0}^{\infty} r^{-j} \varphi(r^j x, r^j y, r^j z, r^j w) < \infty, \\ &\left\|2h\Big(\frac{\mu x + \mu y + zw}{2}\Big) - \mu h(x) - \mu h(y) - h(z)h(w)\right\| \le \varphi(x, y, z, w), \\ &\|h(r^n u^*) - h(r^n u)^*\| \le \varphi(r^n u, r^n u, 0, 0) \end{aligned}$$

for all $\mu \in \mathbb{T}^1$, $u \in \mathcal{U}(\mathcal{A})$, $n = 0, 1, \cdots$, and $x, y, z, w \in \mathcal{A}$. Then the mapping $h : \mathcal{A} \to \mathcal{B}$ is a C^*-algebra homomorphism.

Proof. By a similar method to the proof of Theorem 2.1, one can show that the mapping $h : \mathcal{A} \to \mathcal{B}$ is a C^*-algebra homomorphism.

Corollary 2.6. *Let $h : \mathcal{A} \to \mathcal{B}$ be a mapping satisfying $h(rx) = rh(x)$ for all $x \in \mathcal{A}$ for which there exist constants $\theta \ge 0$ and $p \in [0, 1)$ such that*

$$\begin{aligned} &\left\|2h\Big(\frac{\mu x + \mu y + zw}{2}\Big) - \mu h(x) - \mu h(y) - h(z)h(w)\right\| \\ &\qquad \le \theta(||x||^p + ||y||^p + ||z||^p + ||w||^p), \\ &\|h(r^n u^*) - h(r^n u)^*\| \le 2r^{np}\theta \end{aligned}$$

for all $\mu \in \mathbb{T}^1$, $u \in \mathcal{U}(\mathcal{A})$, $n = 0, 1, \cdots$, and $x, y, z, w \in \mathcal{A}$. Then the mapping $h : \mathcal{A} \to \mathcal{B}$ is a C^*-algebra homomorphism.

Proof. Define $\varphi(x, y, z, w) = \theta(||x||^p + ||y||^p + ||z||^p + ||w||^p)$ and apply Theorem 2.5.

One can obtain similar results to Theorems 2.3 and 2.4 for the Jensen functional equation.

Finally, we are going to investigate C^*-algebra homomorphisms between unital C^*-algebras associated with the Trif functional equation.

Theorem 2.7. *Let $h : \mathcal{A} \to \mathcal{B}$ be a mapping satisfying $h(rx) = rh(x)$ for all $x \in \mathcal{A}$ for which there exists a function $\varphi : \mathcal{A}^{d+2} \to [0, \infty)$ such that*

$$\widetilde{\varphi}(x_1, \cdots, x_d, z, w) := \sum_{j=0}^{\infty} r^{-j} \varphi(r^j x_1, \cdots, r^j x_d, r^j z, r^j w) < \infty,$$

$$\|d\ {}_{d-2}C_{l-2} h\Big(\frac{\mu x_1 + \cdots + \mu x_d}{d} + \frac{zw}{d\ {}_{d-2}C_{l-2}}\Big) + {}_{d-2}C_{l-1} \sum_{j=1}^{d} \mu h(x_j)$$
$$- l \sum_{1 \le j_1 < \cdots < j_l \le d} \mu h\Big(\frac{x_{j_1} + \cdots + x_{j_l}}{l}\Big) - h(z)h(w)\| \le \varphi(x_1, \cdots, x_d, z, w),$$
$$\|h(r^n u^*) - h(r^n u)^*\| \le \varphi(\underbrace{r^n u, \cdots, r^n u}_{d \text{ times}}, 0, 0)$$

for all $\mu \in \mathbb{T}^1$, $u \in \mathcal{U}(\mathcal{A})$, $n = 0, 1, \cdots$, and $x_1, \cdots, x_d, z, w \in \mathcal{A}$. Then the mapping $h : \mathcal{A} \to \mathcal{B}$ is a C^*-algebra homomorphism.

Proof. By a similar method to the proof of Theorem 2.1, one can show that the mapping $h : \mathcal{A} \to \mathcal{B}$ is a C^*-algebra homomorphism.

Corollary 2.8. *Let $h : \mathcal{A} \to \mathcal{B}$ be a mapping satisfying $h(rx) = rh(x)$ for all $x \in \mathcal{A}$ for which there exist constants $\theta \ge 0$ and $p \in [0, 1)$ such that*

$$\Big\|d\ {}_{d-2}C_{l-2} h\Big(\frac{\mu x_1 + \cdots + \mu x_d}{d} + \frac{zw}{d\ {}_{d-2}C_{l-2}}\Big) + {}_{d-2}C_{l-1} \sum_{j=1}^{d} \mu h(x_j)$$
$$- l \sum_{1 \le j_1 < \cdots < j_l \le d} \mu h\Big(\frac{x_{j_1} + \cdots + x_{j_l}}{l}\Big) - h(z)h(w)\Big\|$$
$$\le \theta\Big(\sum_{j=1}^{d} ||x_j||^p + ||z||^p + ||w||^p\Big),$$
$$\|h(r^n u^*) - h(r^n u)^*\| \le d r^{np} \theta$$

for all $\mu \in \mathbb{T}^1$, $u \in \mathcal{U}(\mathcal{A})$, $n = 0, 1, \cdots$, and $x_1, \cdots, x_d, z, w \in \mathcal{A}$. Then the mapping $h : \mathcal{A} \to \mathcal{B}$ is a C^*-algebra homomorphism.

Proof. Define $\varphi(x_1, \cdots, x_d, z, w) = \theta(\sum_{j=1}^{d} ||x_j||^p + ||z||^p + ||w||^p)$ and apply Theorem 2.7.

One can obtain similar results to Theorems 2.3 and 2.4 for the Trif functional equation.

Remark. Replacing $h(z)h(w)$ in each statement of the results obtained above with $h(z)w + zh(w)$, one can show that the mapping $\mathcal{A} \to \mathcal{B}$ is a $\mathbb{C}$-linear $*$-derivation.

References

1. P. Găvruta, *A generalization of the Hyers-Ulam-Rassias stability of approximately additive mappings*, J. Math. Anal. Appl. **184** (1994), 431–436.
2. B. E. Johnson, *Approximately multiplicative maps between Banach algebras*, J. London Math. Soc. (2) **37** (1988), 294–316.
3. K. Jun and Y. Lee, *A generalization of the Hyers-Ulam-Rassias stability of Jensen's equation*, J. Math. Anal. Appl. **238** (1999), 305–315.
4. R. V. Kadison and G. Pedersen, *Means and convex combinations of unitary operators*, Math. Scand. **57** (1985), 249–266.

5. R. V. Kadison and J. R. Ringrose, *Fundamentals of the Theory of Operator Algebras*, Elementary Theory, Academic Press, New York, 1983.
6. C. Park, *On the stability of the linear mapping in Banach modules*, J. Math. Anal. Appl. **275** (2002), 711–720.
7. C. Park, *On an approximate automorphism on a C^*-algebra*, Proc. Amer. Math. Soc. **132** (2004), 1739–1745..
8. Th. M. Rassias, *On the stability of the linear mapping in Banach spaces*, Proc. Amer. Math. Soc. **72** (1978), 297–300.
9. T. Trif, *On the stability of a functional equation deriving from an inequality of Popoviciu for convex functions*, J. Math. Anal. Appl. **272** (2002), 604–616.

In: Inequality Theory and Applications, Volume 4
Editors: Yeol Je Cho et al., pp.119–126
ISBN 1-59454-874-9

ON THE LIN-WONG DIVERGENCE MEASURE OF ENTROPY

C. E. M. PEARCE[1] AND J. PEČARIĆ[2]

ABSTRACT. Some new results are derived relating to the Lin–Wong divergence measure of entropy. Simple proofs are also given for two known results, one of which is refined further.

1. Introduction

A number of measures have been introduced to measure the difference between two probability distributions. Such measures have application *inter alia* to the analysis of contingency tables, approximation of distributions, signal processing and pattern recognition. The best–known such measure is perhaps the I–directed divergence of Kullbach and Leibler [1]. This has the feature of requiring each distribution to be absolutely continuous with respect to the other. More recently Lin and Wong [2] proposed a new information–theoretic measure that does not require this condition.

In this chapter we consider these directed divergences and the relation between them in the basic but practically important case of two probability distributions $p(\cdot)$ and $q(\cdot)$ whose ranges have finite union $\{x_1, \cdots, x_n\}$. For notational convenience we put $p_i := p(x_i)$, $q_i := q(x_i)$. The Kullback and Leibler I– directed divergence between p and q is defined by

$$I(p,q) := \sum_{i=1}^{n} p_i \log \frac{p_i}{q_i}.$$

Throughout the chapter log refers to logarithms taken to base two.

The definition of Kullbach and Leibler suffers the disadvantage of not being symmetric in p and q. A symmetric divergence can, however, be defined by

$$J(p,q) = I(p,q) + I(q,p) = \sum_{i=1}^{n} (p_i - q_i) \log \frac{p_i}{q_i}.$$

Lin and Wong's directed divergence is defined by

$$K(p,q) = \sum_{i=1}^{n} p_i \log \frac{2p_i}{r_i},$$

Received April 20, 2003.
2000 *Mathematics Subject Classification.* 94A17, 26D15.
Key words and phrases. Divergence measure, Jensen inequality, information theory, entropy.
[1] School of Applied Mathematics, University of Adelaide, Adelaide SA 5005, Australia. E-mail: cpearce@maths.adelaide.edu.au
[2] Faculty of Textile Technology, University of Zagreb, Pierottijeva 6, Zagreb, Croatia. Email: pecaric@hazu.hr

where as subsequently $r_i := p_i + q_i$. They have a corresponding symmetric version

$$L(p,q) = K(p,q) + K(q,p) \tag{1.1}$$

and proved the inequality

$$K(p,q) \le \frac{1}{2} I(p,q)$$

between their symmetric directed divergence and that of Kullbach and Leibler. They proved also the following result.

Theorem A. *The Kullbach–Leibler and Lin–Wong divergences satisfy*

$$L(p,q) \le \frac{1}{2} J(p,q).$$

Moreover, if

$$V(p,q) := \sum_{i=1}^{n} |p_i - q_i|$$

denotes the variational distance between the two probability distribution, then we have the following.

Theorem B. *The variational distance and the Lin–Wong symmetric divergence are related through*

$$L(p,q) \le V(p,q).$$

The inequalities above were established by *ad hoc* means. We exhibit machinery for the systematic derivation of such results. We illustrate its use through a number of examples. In Section 2 we provide a simple proof of Theorem A and two upper bounds for K, one involving the differences $p_i - q_i$ and the other the sum r_i. A further upper bound is provided which does not involve logarithms of the probabilities. In Section 3 we derive an upper bound for $V - L$, thus establishing and refining Theorem B. Finally, in Section 4, we provide a rational interpolation between L and V.

Our anvil for forging these results is a recent version of Jensen's inequality established by Matić, Pearce and Pečarić [3]. For discrete–valued measures, this reads as follows.

Theorem C. *Let φ be a convex function on an interval $I \subset \mathbb{R}$ and $\{y_i\}$ $(i = 1, \cdots, n)$ a set of points in I, y_i being associated with the weight $w_i > 0$. If φ is differentiable, then*

$$0 \le \frac{\sum w_i \varphi(y_i)}{\sum w_i} - \varphi\left(\frac{\sum w_i y_i}{\sum w_i}\right)$$

$$\le \frac{\sum w_i y_i \varphi'(y_i)}{\sum w_i} - \frac{1}{(\sum w_i)^2} \sum w_i y_i \sum w_j \varphi'(y_j).$$

This theorem has already proved useful in connection with other questions relating to entropy. Our methodology is to apply Theorem C to judiciously chosen convex functions φ.

2. Theorem A and Bounds for K

We begin with a derivation of Theorem A.

Proof. From its definition, we have

$$L(p,q) = \sum_{i=1}^{n} r_i \left[\frac{p_i}{r_i} \log \frac{2p_i}{r_i} + \frac{q_i}{r_i} \log \frac{2q_i}{r_i}\right], \tag{2.1}$$

and we note that $q_i/r_i = 1 - p_i/r_i$. This suggests the following.

For $0 < t < 1$, define

$$\varphi(t) := t \log(2t) + (1-t)\log[2(1-t)] = 1 + \frac{1}{\ln 2}\left[t \ln t + (1-t)\ln(1-t)\right].$$

Then we have

$$\varphi'(t) = \log \frac{t}{1-t}$$

and

$$\varphi''(t) = \frac{1}{\ln 2}\left(\frac{1}{t} + \frac{1}{1-t}\right) > 0,$$

so φ is convex.

If we apply Theorem C with $w_i = r_i$ and $y_i = p_i/r_i$, the second inequality gives Theorem A immediately.

Theorem 2.1. *We have the upper bounds*

$$0 \le K(p,q) \le \frac{1}{2}\sum_{i=1}^{n}(p_i - q_i)\log\frac{p_i}{r_i} \tag{2.2}$$

and

$$0 \le K(p,q) \le \sum_{i=1}^{n} q_i \log\frac{r_i}{2p_i}. \tag{2.3}$$

Proof. We start with

$$K(p,q) = \sum_{i=1}^{n} p_i \log\frac{2p_i}{r_i} = \sum_{i=1}^{n} r_i \frac{p_i}{r_i}\log\frac{2p_i}{r_i},$$

which suggests we consider

$$\varphi(x) := x\log(2x) = x + x\frac{\ln x}{\ln 2}.$$

Since $\varphi' = 1 + \frac{1}{\ln 2} + \log x$ and $\varphi''(x) = \frac{1}{\ln 2}\frac{1}{x} > 0$, the function φ is convex.

By Theorem C with $w_i = r_i$ and $y_i = p_i/r_i$,

$$\begin{aligned}
0 &\le \frac{1}{2}\sum_{i=1}^{n} r_i \frac{p_i}{r_i}\log\frac{2p_i}{r_i} \\
&\le \frac{1}{2}\sum_{i=1}^{n} p_i\left(1 + \frac{1}{\ln 2} + \log\frac{p_i}{r_i}\right) - \frac{1}{4}\sum r_i\left(1 + \frac{1}{\ln 2} + \log\frac{p_i}{r_i}\right) \\
&= \frac{1}{4}\sum_{i=1}^{n}(p_i - q_i)\left(1 + \frac{1}{\ln 2} + \log\frac{p_i}{r_i}\right) \\
&= \frac{1}{4}\sum_{i=1}^{n}(p_i - q_i)\log\frac{p_i}{r_i}
\end{aligned}$$

and (2.2) is proven. Since

$$\sum_{i=1}^{n}(p_i - q_i)\log\frac{2p_i}{r_i} = \sum_{i=1}^{n}(p_i - q_i)\log\frac{p_i}{r_i},$$

(2.2) becomes

$$0 \le K(p,q) \le \frac{1}{2}K(p,q) - \frac{1}{2}\sum_{i=1}^{n} q_i \log\frac{2p_i}{r_i}$$

or

$$0 \le K(p,q) \le -\sum_{i=1}^{n} q_i \log\frac{2p_i}{r_i},$$

which is equivalent to (2.3).

Remark 2.2. From (2.2) we have also

$$0 \le K(q,p) \le \frac{1}{2}\sum_{i=1}^{n}(q_i - p_i)\log\frac{2q_i}{r_i}. \tag{2.4}$$

From (1.1), (2.2) and (2.4) we get

$$0 \le L(p,q) \le \frac{1}{2}\sum(p_i - q_i)\log\frac{p_i}{q_i} = \frac{1}{2}J(p,q),$$

providing a further derivation of Theorem A.

To illustrate our method, we show how alternative upper bounds for K and so L can be derived directly which do not involve logarithms of probabilities.

Theorem 2.3. *We have the upper bounds*

$$0 \le K(p,q) \le \frac{1}{\ln 2}\sum_{i=1}^{n}(p_i - q_i)\frac{p_i}{r_i} \tag{2.5}$$

and

$$0 \le L(p,q) \le \frac{1}{\ln 2}\sum_{i=1}^{n}\frac{(p_i - q_i)^2}{r_i}. \tag{2.6}$$

Proof. Recasting the definition of K as

$$K(p,q) = \sum_{i=1}^{n} p_i \log\frac{2}{1 + q_i/p_i}$$

suggests we work from

$$\varphi(x) = \log\frac{2}{1+x} = 1 - \frac{\ln(1+x)}{\ln 2}.$$

Since $\varphi'(x) = -\dfrac{1}{\ln 2}\dfrac{1}{1+x}$ and $\varphi'' = \dfrac{1}{\ln 2}\dfrac{1}{(1+x)^2} > 0$, the function φ is convex. Applying Theorem C with $w_i = p_i$ and $y_i = q_i/p_i$ yields

$$\begin{aligned}0 &\le \sum_{i=1}^{n} p_i \log \frac{2p_i}{r_i} \\ &\le -\sum_{i=1}^{n} q_i \frac{1}{\ln 2}\frac{p_i}{r_i} + \sum_{i=1}^{n} p_i \frac{1}{\ln 2}\frac{p_i}{r_i} \\ &= \frac{1}{\ln 2}\sum_{i=1}^{n}(p_i - q_i)\frac{p_i}{r_i}\end{aligned}$$

and (2.5) is proven.

By (2.5), we have also

$$0 \le K(q,p) \le \frac{1}{\ln 2}\sum_{i=1}^{n}(q_i - p_i)\frac{q_i}{r_i} = -\frac{1}{\ln 2}\sum_{i=1}^{n}(p_i - q_i)\frac{q_i}{r_i}.$$

By (1.1), relation (2.6) follows by combining the last inequality with (2.5).

3. Theorem B and Bounds for $V - L$

Theorem 3.1. *The difference between V and L satisfies*

$$0 \le V(p,q) - L(p,q) \le 2\log 5 - 4.$$

Proof. Relation (2.1) suggests the possibility of working from

$$\begin{aligned}\varphi(t) &:= 2t - 1 + t\log(2t) + (1-t)\log[2(1-t)] \\ &= 2t + \frac{1}{\ln 2}(t\ln t + (1-t)\ln(1-t))\end{aligned}$$

for $t \in [0, 1/2]$. We have

$$\varphi'(t) = 2 + \frac{1}{\ln 2}(\ln t - \ln(1-t)) \quad \text{and} \quad \varphi''(t) = \frac{1}{\ln 2}\left(\frac{1}{t} + \frac{1}{1-t}\right) > 0.$$

Thus φ is strictly convex and since $\varphi(0+) = 0$ and $\varphi(1/2) = 0$, we have that $\varphi(t) \le 0$ with equality only for $t = 0, 1/2$. A unique minimum is achieved by φ where $\varphi'(t_0) = 0$, that is,

$$2 + \frac{1}{\ln 2}\ln\frac{t_0}{1-t_0} = 0,$$

whence $t_0 = 1/5$.

Since $\varphi(1/5) = 2 - \log 5$, we have

$$2 - \log 5 \le 2t - 1 + t\log(2t) + (1-t)\log[2(1-t)] \le 0.$$

By symmetry, we have for $t \in [0,1]$ that

$$|(1-t) - t| + 2 - \log 5 \le t\log(2t) + (1-t)\log(2(t-1)) \le |(1-t) - t|, \tag{3.1}$$

with equality on the right only for $t = 0, 1/2, 1$.

Set $t = p_i/r_i$. Multiplication by r_i and summation provides

$$\sum_{i=1}^{n} r_i \left| \frac{p_i}{r_i} - \frac{q_i}{r_i} \right| + 2(2 - \log 5) \le L(p,q) \le \sum_{i=1}^{n} r_i \left| \frac{p_i}{r_i} - \frac{q_i}{r_i} \right|,$$

that is,

$$V(p,q) - 2(\log 5 - 2) \le L(p,q) \le V(p,q),$$

whence the required result.

4. Interpolation between L and V

To conclude, we provide a rational interpolation between L and V. Our development combines results established earlier in the paper with Lemma 4.1 below.

We have two rational upper bounds for L, inequality (3.1) stating

$$L(p,q) \le V(p,q) = \sum_{i=1}^{n} |p_i - q_i|$$

while (2.6) provides

$$L(p,q) \le \frac{1}{\ln 2} \sum_{i=1}^{n} \frac{(p_i - q_i)^2}{r_i}.$$

How do these compare?

The condition that

$$\frac{1}{\ln 2} \frac{|p_i - q_i|}{r_i} \le 1$$

is readily seen to be that

$$\frac{1 - \ln 2}{1 + \ln 2} \le \frac{q_i}{p_i} \le \frac{1 + \ln 2}{1 - \ln 2}. \tag{4.1}$$

If this holds for each i, then we have that

$$0 \le L(p,q) \le \frac{1}{\ln 2} \sum_{i=1}^{n} \frac{(p_i - q_i)^2}{r_i} \le \sum_{i=1}^{n} |p_i - q_i| \equiv V(p,q).$$

On the other hand, if (4.1) fails for every i then

$$L(p,q) \le V(p,q) < \frac{1}{\ln 2} \sum_{i=1}^{n} \frac{(p_i - q_i)^2}{r_i}.$$

This raises the question as to whether we can modify the far right of (2.6) so as to interpolate between L and V. A natural candidate for interpolation is the choice $\sum_{i=1}^{n} \alpha_i |p_i - q_i|$, where

$$\alpha_i := \min\left\{1, \frac{1}{\ln 2} \frac{|p_i - q_i|}{r_i}\right\}.$$

Our final result establishes its adequacy.

Lemma 4.1. *The inequality*

$$t \ln 2t + (1-t)\ln 2(1-t) \; - (2t-1)^2 \leq 0 \tag{4.2}$$

holds for $t \in [0,1]$, *with equality only at* $t = 1/2$.

Proof. For $t \in [0, 1/2]$, set

$$f(t) := t \ln 2t + (1-t)\ln 2(1-t) \; - (2t-1)^2.$$

Then

$$\begin{aligned} f'(t) &= \ln 2t - \ln 2(1-t) \; - 4(2t-1), \\ f''(t) &= t^{-1} + (1-t)^{-1} - 8, \\ f'''(t) &= (1-t)^{-2} - t^{-2}. \end{aligned}$$

Hence $f'''(t) < 0$ on $(0, 1/2)$ and so f' is strictly concave on $[0, 1/2]$. Also $f''(0+) > 0$, $f''(1/2) < 0$, while $f'(0+) < 0$, $f'(1/2) = 0$. Therefore $f'(t) < 0$ for $t \in [0, t_1)$ and $f'(t) > 0$ for $t \in (t_1, 1/2]$, where t_1 is the unique solution of $f'(t_1) = 0$ on $(0, 1/2)$.

We deduce that f is strictly decreasing for $t \in [0, t_1)$ and strictly increasing for $t \in (t_1, 1/2)$. Since $f(0) = \ln 2 - 1 < 0$ and $f(1/2) = 0$, we conclude that $f(t) \leq 0$ on $[0, 1/2]$ with equality only for $t = 1/2$. The lemma follows, as the left–hand side of (4.2) is symmetric about $t = 1/2$ for $t \in [0, 1]$.

Corollary 4.2. *For* $t \in [0,1]$,

$$t \ln 2t + (1-t)\ln 2(1-t) \; \leq \min\left\{ |2t-1|, \frac{(2t-1)^2}{\ln 2} \right\}$$

with equality only for $t = 0, 1/2, 1$.

Proof. This follows by combining the preceding lemma with the second part of inequality (3.1).

Theorem 4.3. *The entropy divergence measures* L *and* V *are rationally interpolated by*

$$L(p,q) \leq \sum_{i=1}^{n} \alpha_i |p_i - q_i| \leq V(p,q).$$

Proof. Set $t = p_i/q_i$ in Corollary 4.2. Multiplication by r_i and summation provide

$$L(p,q) \leq \sum_{i=1}^{n} \alpha_i |p_i - q_i|.$$

On the other hand,

$$\sum_{i=1}^{n} \alpha_i |p_i - q_i| \leq \sum_{i=1}^{n} |p_i - q_i| = V(p,q),$$

giving the desired result.

Remark 4.4. The first inequality in Theorem 4.3 becomes an equality only when each p_i/r_i assumes one of the values $0, 1/2, 1$. The first and last of these possibilities correspond to one of the values p_i, q_i being positive and the other zero, the intermediate possibility to $p_i = q_i$. That is, the two distributions have to agree on each value x_i of their domain on which both occur with positive probability. This allows the extreme case when the supports of the two distributions are disjoint.

The second inequality becomes an equality only when $\alpha_i = 1$ for each i, that is, (4.1) holds for each i, or every ratio q_i/p_i is sufficiently close to unity.

Thus both inequalities simultaneously become equalities (and so $V = L$) only when $p_i = q_i$ for all $i = 1, \cdots, n$.

References

1. S. Kullback and R. A. Leibler, *On information and sufficiency*, Ann. Math. Statist. **22** (1951), 79–86.
2. J. Lin and S. K. M. Wong, *A new directed divergence measure and its characterization*, Int. J. General Systems **17** (1989), 73–81.
3. M. Matić, C. E. M. Pearce and J. Pečarić, *Some refinements of Shannon's inequalities*, ANZIAM J. **43** (2002), 493–511.

In: Inequality Theory and Applications, Volume 4
Editors: Yeol Je Cho et al., pp.127–135

ISBN 1-59454-874-9

ALGEBRAICALLY DETERMINED p-SERIES

A. Sofo[1], S. S. Dragomir[1] and Y. J. Cho[2]

Abstract. We determine closed form representations of quadratic p-series from a recursion point of view.

1. Introduction

A p-series (or Riemann zeta series) is defined as

$$S(p) := \sum_{n=1}^{\infty} \frac{1}{n^p}, \tag{1.1}$$

where p is a positive integer bigger than or equal to one.

Similarly, an alternating p-series (or Dirichlet series) is defined as

$$D(p) := \sum_{n=1}^{\infty} \frac{(-1)^{n+1}}{n^p}. \tag{1.2}$$

The alternating series (1.2) is associated with (1.1) by the following identity

$$D(p) = \left(1 - \frac{1}{2^{p-1}}\right) S(p) \tag{1.3}$$

and, for p even, we can express (1.1) and (1.2) in closed form. Table 1 displays some results.

In this short note, we will algebraically associate (1.1) with

$$R(p) := \sum_{n=1}^{\infty} \frac{1}{(2n-1)^p}, \tag{1.4}$$

$$M(p) := \sum_{n=1}^{\infty} \frac{1}{(2n+1)^p}, \tag{1.5}$$

$$T(p) := \sum_{n=1}^{\infty} \frac{1}{(4n^2-1)^p}. \tag{1.6}$$

Received January 3, 2004.
2000 *Mathematics Subject Classification.* Primary 40C99; Secondary 11M35.
Key Words and Phrases. p-Series, recursion, Closed form.

[1] School of Computer Science and Mathematics, Victoria University of Technology, PO Box 14428, MCMC 8001, Victoria, Australia. E-mails: sofo@csm.vu.edu.au (http://rgmia.vu.edu. au/sofo) and Sever.Dragomir@vu.edu.au (http://rgmia.vu.edu.au/SSDragomirWeb.html)

[2] Department of Mathematics Education, Gyeongsang National University, Chinju 660-701, Korea. E-mail: yjcho@nongae.gsnu.ac.kr

Similarly, we will consider and associate algebraically the following

(1.7) $$Q(p) := \sum_{n=1}^{\infty} \frac{(-1)^{n+1}}{(2n-1)^p},$$

(1.8) $$V(p) := \sum_{n=1}^{\infty} \frac{(-1)^{n+1}}{(2n+1)^p},$$

(1.9) $$W(p) := \sum_{n=1}^{\infty} \frac{(-1)^{n+1}}{(4n^2-1)^p}.$$

We now consider the non-alternating case.

2. Non-Alternating Case

The following lemma will be required later.

Lemma 1.

(2.1) (i) $$\sum_{n=1}^{\infty} \frac{1}{(2n-1)^p} = \sum_{n=1}^{\infty} \frac{1}{n^p} - \sum_{n=1}^{\infty} \frac{1}{(2n)^p};$$

(2.2) (ii) $$\sum_{n=1}^{\infty} \left\{ \frac{(2n+1)^p - (2n-1)^p}{(4n^2-1)^p} \right\} = 1;$$

(2.3) (iii) $$\sum_{n=1}^{\infty} \frac{1}{(2n+1)^p} = \left(1 - \frac{1}{2^p}\right) \sum_{n=1}^{\infty} \frac{1}{n^p} - 1.$$

Proof. (i) Follows directly upon subtracting $\sum_{n=1}^{\infty} \frac{1}{(2n)^p}$ from (1.1).
(ii) Follows from the subtraction of (1.4) and (1.5).
(iii) From (2.1), we have

(2.4) $$\sum_{n=1}^{\infty} \frac{1}{(2n+1)^p} = -1 + \sum_{n=1}^{\infty} \frac{1}{(2n-1)^p}$$

and, utilizing (2.2),

$$\sum_{n=1}^{\infty} \frac{1}{(2n+1)^p} = \sum_{n=1}^{\infty} \frac{1}{n^p} - \sum_{n=1}^{\infty} \frac{1}{(2n)^p} - 1.$$

Upon rearrangement (2.3) follows.

The following lemma will be used later.

Lemma 2. *For* $p = 1, 2, \cdots$ *if*

(2.5) $$A_j = \lim_{x \to \frac{1}{2}} \frac{1}{(j-1)! 2^{p+j-1}} \frac{d^{j-1}}{dx^{j-1}} \left[\left(x - \frac{1}{2}\right)^p F(x) \right], \quad j = 1, 2, \cdots, p,$$

where $F(x) = \frac{1}{\left(x^2 - \frac{1}{4}\right)^p}$, *then* $\sum_{j=1}^{p} |A_j| = \frac{1}{2}$.

Proof. For

$$j = 1; \quad |A_1| = \frac{1}{0!} \frac{1}{2^p};$$

$$j = 2; \quad |A_j| = \frac{1}{1!} \frac{p}{2^{p+1}};$$

$$\vdots$$

$$j = p; \quad |A_p| = \frac{p(p+1) \cdots (p+p-2)}{(p-1)! 2^{p+p-1}},$$

then

$$\sum_{j=1}^{p} |A_j| = \frac{1}{0!}\frac{1}{2^p} + \frac{p}{1!2^{p+1}} + \cdots + \frac{p(p+1)\cdots(p+p-2)}{(p-1)!2^{p+p-1}}$$
$$= \sum_{j=1}^{p} \frac{(p+j-2)!}{(j-1)!2^{p+j-1}(p-1)!}$$
$$= \sum_{j=1}^{p} \frac{1}{2^{p+j-1}} \binom{p+j-2}{j-1} = \frac{1}{2^p} \cdot 2^{p-1} = \frac{1}{2}$$

and hence the lemma is proved.

We now state the main theorem for the non-alternating case.

Theorem 1. *For p a positive integer,*

$$\sum_{n=1}^{\infty} \frac{1}{(4n^2-1)^p} \tag{2.6}$$
$$= \frac{(-1)^{p+1}}{2} + (-1)^p \sum_{k=1}^{p} |A_{p-k+1}| \left(1+(-1)^k\right)\left(1-\frac{1}{2^k}\right) S(k),$$

where A_j is defined by (2.5) and $S(k)$ by (1.1).

Proof.

$$\sum_{n=1}^{\infty} \frac{1}{(4n^2-1)^p} = \sum_{n=1}^{\infty}\sum_{j=1}^{p} \left\{ \frac{A_j}{(2n-1)^{p-j+1}} + \frac{B_j}{(2n+1)^{p-j+1}} \right\},$$

where A_j is given by (2.5) and, similarly,

$$B_j = \lim_{x \to \frac{-1}{2}} \frac{1}{(j-1)!} \frac{d^{j-1}}{dx^{j-1}} \left[\frac{\left(x+\frac{1}{2}\right)^p F(x)}{2^{p+j-1}} \right]; \quad j = 1, 2, \cdots, p. \tag{2.7}$$

Since A_j and B_j are related, by$A_j = (-1)^{j+1}|A_j|$, and $B_j = (-1)^p |A_j|$, we can write

$$\sum_{n=1}^{\infty} \frac{1}{(4n^2-1)^p} = \sum_{n=1}^{\infty}\sum_{j=1}^{p} \left\{ \frac{(-1)^{j+1}|A_j|}{(2n-1)^{p-j+1}} + \frac{(-1)^p|A_j|}{(2n+1)^{p-j+1}} \right\}$$
$$= \sum_{j=1}^{p} |A_j| \sum_{n=1}^{\infty} \left\{ \frac{(-1)^{j+1}}{(2n-1)^{p-j+1}} + \frac{(-1)^p}{(2n+1)^{p-j+1}} \right\}$$

upon interchanging the summation. Using (2.4), we have

$$\sum_{n=1}^{\infty} \frac{1}{(4n^2-1)^p} = \sum_{j=1}^{p} |A_j| \sum_{n=1}^{\infty} \left\{ \frac{(-1)^{j+1}+(-1)^p}{(2n-1)^{p-j+1}} - (-1)^p \right\}$$
$$= \sum_{j=1}^{p} (-1)^{p+1}|A_j| + \sum_{j=1}^{p} |A_j| \sum_{n=1}^{\infty} \left\{ \frac{(-1)^{j+1}+(-1)^p}{(2n-1)^{p-j+1}} \right\}.$$

From (2.2), it follows that

$$\sum_{n=1}^{\infty} \left\{ \frac{1}{(2n-1)^{p-j+1}} \right\} = \left(1 - \frac{1}{2^{p-j+1}}\right) \sum_{n=1}^{\infty} \frac{1}{n^{p-j+1}},$$

so that

$$\sum_{n=1}^{\infty} \frac{1}{(4n^2-1)^p}$$
$$= \sum_{j=1}^{p} (-1)^{p+1} |A_j| + \sum_{j=1}^{p} |A_j| \sum_{n=1}^{\infty} \left\{ (-1)^{j+1} + (-1)^p \right\} \left(1 - \frac{1}{2^{p-j+1}}\right) \frac{1}{n^{p-j+1}}$$
$$= (-1)^{p+1} \sum_{j=1}^{p} |A_j| + \sum_{j=1}^{p} |A_j| \left((-1)^{j+1} + (-1)^p \right) \left(1 - \frac{1}{2^{p-j+1}}\right) \sum_{n=1}^{\infty} \frac{1}{n^{p-j+1}}.$$

Now, using Lemma 2, we have

$$\sum_{n=1}^{\infty} \frac{1}{(4n^2-1)^p}$$
$$= \frac{(-1)^{p+1}}{2} + \sum_{j=1}^{p} |A_j| \left((-1)^{j+1} + (-1)^p \right) \left(1 - \frac{1}{2^{p-j+1}}\right) \sum_{n=1}^{\infty} \frac{1}{n^{p-j+1}},$$

and, by the change of counter $k = p - j + 1$, we arrive at our result (2.6) and hence the theorem is proved.

Table 1 lists some values of (2.2) and (2.6).

TABLE 1. Some values of $S(p)$, $R(p)$ and $T(p)$

p	$S(p)$; (1.1)	$R(p)$; (1.4)	$T(p)$; (1.6)
1	—	—	$\frac{1}{2}$
2	$\frac{\pi^2}{6}$	$\frac{\pi^2}{8}$	$\frac{\pi^2}{16} - \frac{1}{2}$
3	—	—	$\frac{1}{2} - \frac{3\pi^2}{64}$
4	$\frac{\pi^4}{90}$	$\frac{\pi^4}{96}$	$\frac{\pi^4+30\pi^2-384}{768}$
5	—	—	$\frac{1536-105\pi^2-5\pi^4}{2^{10}\cdot 3}$
6	$\frac{\pi^6}{945}$	$\frac{\pi^6}{960}$	$\frac{2\pi^6+105\pi^4+1890\pi^2-30720}{2^{12}\cdot 3\cdot 5}$
7	—	—	$\frac{2^{12}\cdot 3\cdot 5-3465\pi^2-210\pi^4-7\pi^6}{2^{13}\cdot 3\cdot 5}$
8	$\frac{\pi^8}{9450}$	$\frac{17\pi^8}{161280}$	$\frac{17\pi^8+1512\pi^6+34650\pi^4+540540\pi^2-2^{15}\cdot 3^2\cdot 5\cdot 7}{2^{16}\cdot 3^2\cdot 5\cdot 7}$
9	—	—	$\frac{2^{17}\cdot 5\cdot 7-225225\pi^2-15015\pi^4-770\pi^6-17\pi^8}{2^{18}\cdot 5\cdot 7}$
10	$\frac{\pi^{10}}{93555}$	$\frac{1023\pi^{10}}{2^{10}\cdot 93555}$	$\frac{62\pi^{10}+8415\pi^8+270270\pi^6+4729725\pi^4+68918850\pi^2-2^{19}\cdot 3^4\cdot 5\cdot 7}{2^{16}\cdot 3^2\cdot 5\cdot 7}$

Jolley [4] lists the values of $T(1)$, $T(2)$ and $T(3)$ which he attributes to Adams [1].
Jolley also lists $\sum_{n=1}^{\infty} \frac{n}{(4n^2-1)^2} = \frac{1}{8}$, which follows directly from (2.1) for the case $p = 2$ and, similarly, for $p = 3$, from (2.1) we have

$$\sum_{n=1}^{\infty} \frac{n^2}{(4n^2-1)^3} = \frac{\pi^2}{256}.$$

We now consider the alternating case.

3. Alternating Case

The following lemma will be useful.

Lemma 3. *For $p = 1, 2, 3, \cdots$,*

$$\sum_{n=1}^{\infty} (-1)^{n+1} \left(\frac{(2n+1)^p - (2n-1)^p}{(4n^2-1)^p} \right) = 1$$

or

$$\sum_{n=1}^{\infty} \frac{(-1)^{n+1}}{(2n+1)^p} = 1 - \sum_{n=1}^{\infty} \frac{(-1)^{n+1}}{(2n-1)^p}. \tag{3.1}$$

The proof of Lemma 3 is relatively straightforward and will not be given here. The following theorem applies.

Theorem 2. *For p a positive integer,*

$$\sum_{n=1}^{\infty} \frac{(-1)^{n+1}}{(4n^2-1)^p} = \frac{(-1)^p}{2} + (-1)^{p+1} \sum_{k=1}^{p} |A_{p-k+1}| \left(1 - (-1)^k\right) Q(k), \tag{3.2}$$

where A_j is given by (2.5) and $Q(k)$ by (1.7).

Proof.

$$\sum_{n=1}^{\infty} \frac{(-1)^{n+1}}{(4n^2-1)^p} = \sum_{n=1}^{\infty} (-1)^{n+1} \sum_{j=1}^{p} \left\{ \frac{A_j}{(2n-1)^{p-j+1}} + \frac{B_j}{(2n+1)^{p-j+1}} \right\},$$

where A_j is given by (2.5) and B_j by (2.7). Since A_j and B_j are related, we can write

$$\begin{aligned} \sum_{n=1}^{\infty} \frac{(-1)^{n+1}}{(4n^2-1)^p} &= \sum_{n=1}^{\infty} (-1)^{n+1} \sum_{j=1}^{p} |A_j| \left\{ \frac{(-1)^{j+1}}{(2n-1)^{p-j+1}} + \frac{(-1)^p}{(2n+1)^{p-j+1}} \right\} \\ &= \sum_{j=1}^{p} |A_j| \sum_{n=1}^{\infty} (-1)^{n+1} \left\{ \frac{(-1)^{j+1}}{(2n-1)^{p-j+1}} + \frac{(-1)^p}{(2n+1)^{p-j+1}} \right\} \end{aligned}$$

upon interchanging the summation. Using (3.1), we may write

$$\begin{aligned} &\sum_{n=1}^{\infty} \frac{(-1)^{n+1}}{(4n^2-1)^p} \\ &= \sum_{j=1}^{p} |A_j| \sum_{n=1}^{\infty} \left\{ (-1)^p + \frac{(-1)^{n+1}}{(2n-1)^{p-j+1}} \left((-1)^{j+1} - (-1)^p \right) \right\} \\ &= \sum_{j=1}^{p} (-1)^p |A_j| + \sum_{j=1}^{p} |A_j| \left((-1)^{j+1} - (-1)^p \right) \sum_{n=1}^{\infty} \frac{(-1)^{n+1}}{(2n-1)^{p-j+1}}. \end{aligned}$$

Now, using Lemma 2, we have

$$\sum_{n=1}^{\infty} \frac{(-1)^{n+1}}{(4n^2-1)^p} = \frac{(-1)^p}{2} + \sum_{j=1}^{p} |A_j| \left((-1)^{j+1} - (-1)^p \right) \sum_{n=1}^{\infty} \frac{(-1)^{n+1}}{(2n-1)^{p-j+1}},$$

and, by the change of counter $k = p - j + 1$, we arrive at our result (3.2) and hence the theorem is proved.

The following Table 2 lists some values of (1.7) and (3.2).

TABLE 2. Some values of (1.7) and (3.2).

p	$Q(p)$; (1.7)	$W(p)$; (2.2)
1	$\frac{\pi}{4}$	$\frac{\pi}{4}-\frac{1}{2}$
2	—	$\frac{1}{2}-\frac{\pi}{8}$
3	$\frac{\pi^3}{32}$	$\frac{\pi^3+12\pi-64}{2^7}$
4	—	$\frac{64-10\pi-\pi^3}{2^7}$
5	$\frac{5\pi^5}{1536}$	$\frac{5\pi^5+180\pi^3+1680\pi-2^{12}\cdot 3}{2^{13}\cdot 3}$
6	—	$\frac{2^{13}-1008\pi-112\pi^3-5\pi^5}{2^{14}}$
7	$\frac{61\pi^7}{184320}$	$\frac{61\pi^7+4200\pi^5+75600\pi^3+665280\pi-2^{17}\cdot 3^2\cdot 5}{2^{18}\cdot 3^2\cdot 5}$
8	—	$\frac{2^{16}\cdot 3^2\cdot 5-308880\pi-35640\pi^3-2250\pi^5-61\pi^7}{2^{17}\cdot 3^2\cdot 5}$
9	$\frac{277\pi^9}{8257536}$	$\frac{277\pi^9+30744\pi^7+831600\pi^5+12108096\pi^3+103783680-2^{24}\cdot 3^2\cdot 7}{2^{25}\cdot 3^2\cdot 7}$
10	—	$\frac{2^{25}\cdot 3^2\cdot 7-196053840\pi-23063040\pi^3-1681680\pi^5-75152\pi^7-1385\pi^9}{2^{26}\cdot 3^2\cdot 7}$

Jolley [4] lists the value of $W(1)$, from which we can also obtain

$$\sum_{n=1}^{\infty} \frac{(-1)^{n+1} n}{4n^2-1} = \frac{1}{4}.$$

For $p = 2$ and $p = 4$, we can obtain, respectively,

$$\sum_{n=1}^{\infty} \frac{(-1)^{n+1} n^2}{(4n^2-1)^2} = \frac{\pi}{32};$$

$$\sum_{n=1}^{\infty} \frac{(-1)^{n+1} n^2}{(4n^2-1)^4} = \frac{\pi}{256};$$

$$\sum_{n=1}^{\infty} \frac{(-1)^{n+1} n^4}{(4n^2-1)^4} = \frac{\pi^3-2\pi}{2048}.$$

Some particular cases of (3.2) are also given by Gradshteyn and Ryzhik [3]. We can now make the following remarks.

Remark 1. Some of the series (1.4), (1.6) or (1.9) can be determined in closed form by Fourier series analysis.

The Fourier series representation

$$\frac{1}{2}-\frac{\pi}{4}\sin x=\sum_{n=1}^{\infty}\frac{\cos 2nx}{(4n^2-1)},\qquad x\in\left[0,\frac{\pi}{2}\right],$$

leads to the result (1.6) and (1.9) for $p=1$.

Similarly, from

$$\frac{\pi}{4}\left(\frac{\pi}{2}-|x|\right)=\sum_{n=1}^{\infty}\frac{\cos(2n-1)x}{(2n-1)^2},\qquad x\in[-\pi,\pi],$$

we can obtain (1.4) for $p=2$.

Luo, Guo and Qi [6] have used similar Fourier series analysis to obtain recurrence formulas for the evaluation of the Riemann zeta series (1.1).

Remark 2. Some of the series (1.4), (1.6) or (1.9) can also be determined from contour integral methods and the reader is referred to the excellent paper by Flajolet and Salvy [2].

Remark 3. The series (1.4), (1.6) and (1.9) can also be represented as a generalised hypergeometric function.

Consider the left hand side of (3.2)

$$\sum_{n=0}^{\infty}\frac{(-1)^{n+1}}{(4n^2-1)^p}=(-1)^{p-1}+\sum_{n=1}^{\infty}\frac{(-1)^{n+1}}{(4n^2-1)^p}.$$

Let $T_n=\frac{(-1)^{n+1}}{(4n^2-1)^p}$, where $T_0=(-1)^{p+1}$, and hence

$$\frac{T_{n+1}}{T_n}=\frac{(-1)\left(n-\frac{1}{2}\right)^p(n+1)}{(n+1)\left(n+\frac{3}{2}\right)^p}.$$

We can now write

$$\sum_{n=1}^{\infty}\frac{(-1)^{n+1}}{(4n^2-1)^p}=(-1)^p+(-1)^{p+1}\ {}_{p+1}F_p\left[\begin{matrix}1,-\frac{1}{2},-\frac{1}{2},\ldots,-\frac{1}{2}\\ \frac{3}{2},\frac{3}{2},\ldots,\frac{3}{2}\end{matrix}\middle|-1\right]\tag{3.3}$$

or, alternatively,

$$\begin{aligned}&{}_{p+1}F_p\left[\begin{matrix}1,-\frac{1}{2},-\frac{1}{2},\ldots,-\frac{1}{2}\\ \frac{3}{2},\frac{3}{2},\ldots,\frac{3}{2}\end{matrix}\middle|-1\right]\\ &=1-\sum_{k=1}^{p}|A_{p-k+1}|+\sum_{k=1}^{p}\left(1-(-1)^k\right)|A_{p-k+1}|\sum_{n=1}^{\infty}\frac{(-1)^{n+1}}{(2n-1)^k}\\ &=\frac{1}{2}+\sum_{k=1}^{p}|A_{p-k+1}|\left(1-(-1)^k\right)Q(k),\end{aligned}$$

where A_j is defined by (2.5).

Remark 4. Similar analysis to the above can be done for the arbitrarily shifted denominator of the series $\sum_{n=1}^{\infty} \frac{1}{(4n^2-a^2)^p}$, where the $(2n-a)$ term is excluded.

Remark 5. Either of the series (1.6) or (1.9) can be expressed in terms of the Lerch transcedent or the Catalan beta function. In particular, from (1.9),

$$\sum_{n=1}^{\infty} \frac{(-1)^{n+1}}{(4n^2-1)^p} = \sum_{n=0}^{\infty} \frac{(-1)^{n+1}}{(4n^2-1)^p} + (-1)^p. \tag{3.4}$$

If we define the Lerch transcedent, $\phi(z,s,a)$, as

$$\phi(z,s,a) = \sum_{n=0}^{\infty} \frac{z^n}{(n+a)^s}, \tag{3.5}$$

where the $n+a=0$ term is excluded, we see that, in particular, the Catalan beta function

$$\beta(s) = \sum_{n=0}^{\infty} \frac{(-1)^n}{(2n+1)^s} = 2^{-s}\phi\left(-1,s,\frac{1}{2}\right) \tag{3.6}$$

and the Catalan constant, for $s=2$,

$$\beta(2) = C = \sum_{n=0}^{\infty} \frac{(-1)^n}{(2n+1)^2} = \frac{1}{4}\phi\left(-1,2,\frac{1}{2}\right) \sim 0.91596.$$

The Lerch transcedent is also related to the generalised zeta function $\zeta(s,a)$ and is defined by

$$\zeta(s,a) = \phi(1,s,\alpha) = \sum_{n=0}^{\infty} \frac{1}{(n+\alpha)^s},$$

$$\mathbb{R}(s) > 1,\ \alpha \in C|Z_0^-;\ \ Z_0^- := Z^- \cup \{0\};\ \ Z^- := \{-1,-2,-3,\cdots\}.$$

For more properties of the generalised zeta function and closed form representations of series involving zeta functions, the reader is referred to the excellent article of Lee and Choi [5].

From (3.4), (3.5) and (3.6), we have

$$\begin{aligned}
&\sum_{n=1}^{\infty} \frac{(-1)^{n+1}}{(4n^2-1)^p} \\
&= (-1)^p + \sum_{n=0}^{\infty} (-1)^{n+1} \sum_{j=1}^{p} \left\{ \frac{A_j}{(2n-1)^{p-j+1}} + \frac{B_j}{(2n+1)^{p-j+1}} \right\} \\
&= (-1)^p - \sum_{j=1}^{p}\sum_{n=0}^{\infty} \left\{ \frac{(-1)^n A_j}{(2n-1)^{p-j+1}} + \frac{(-1)^n B_j}{(2n+1)^{p-j+1}} \right\} \\
&= (-1)^p - \sum_{j=1}^{p} (2)^{-(p-j+1)} \left\{ A_j\phi\left(-1,p-j+1,-\frac{1}{2}\right) \right. \\
&\quad \left. + B_j\phi\left(-1,p-j+1,\frac{1}{2}\right) \right\} \\
&= (-1)^p - \sum_{j=1}^{p} \left\{ A_j (2)^{-(p-j+1)} \phi\left(-1,p-j+1,-\frac{1}{2}\right) + B_j\beta(p-j+1) \right\}.
\end{aligned}$$

Using the relationship between A_j and B_j, we have

$$\begin{aligned}(3.7)\qquad &\sum_{n=1}^{\infty}\frac{(-1)^{n+1}}{(4n^2-1)^p}\\ &=(-1)^p-\sum_{j=1}^{p}|A_j|\left\{(-1)^{j+1}\,2^{-(p-j+1)}\phi\left(-1,p-j+1,-\frac{1}{2}\right)\right.\\ &\quad\left.+(-1)^p\,\beta\,(p-j+1)\right\}\end{aligned}$$

and, by the change of counter $k=p-j+1$, we have

$$\sum_{n=1}^{\infty}\frac{(-1)^{n+1}}{(4n^2-1)^p}=(-1)^p-(-1)^p\sum_{j=1}^{p}|A_{p-j+1}|\left\{\left(-\frac{1}{2}\right)^k\phi\left(-1,k,-\frac{1}{2}\right)+\beta\,(k)\right\}.$$

From (3.3) and (3.7), we also see that

$$\begin{aligned}&{}_{p+1}F_p\left[\begin{matrix}1,-\frac{1}{2},-\frac{1}{2},\cdots,-\frac{1}{2}\\ \frac{3}{2},\frac{3}{2},\cdots,\frac{3}{2}\end{matrix}\middle|-1\right]\\ &=\sum_{j=1}^{p}|A_{p-j+1}|\left\{\left(-\frac{1}{2}\right)^k\phi\left(-1,k,-\frac{1}{2}\right)+\beta\,(k)\right\}\\ &=\frac{1}{2}+\sum_{k=1}^{p}|A_{p-j+1}|\left(1-(-1)^k\right)Q(k).\end{aligned}$$

References

1. E. P. Adams, *Smithsonian mathematical Formulae*, Washington, Smithsonian Institute (1922.).
2. P. Flajolet and B. Salvy, *Euler sums and contour integral representations*, Experimental Mathematics **7** (1998), 15–35.
3. I. S. Gradshteyn and I. M. Ryzhik, *Table of Integrals, Series and Products*, Academic Press, New York, 1994.
4. L. B. W. Jolley, *Summation of Series, 2nd Edition*, Dover Publications, Inc., New York, 1961.
5. Q. M. Luo, B. N. Guo and F. Qi, *On evaluation of the Riemann Zeta Function $\zeta(s)$*, RGMIA Res. Rep. Coll. **6(1)** (2003), [Online: http://rgmia.vu.edu. au/v6n1.html].
6. H. Lee and J. Choi, *Series associated with the Zeta function*, Inequalities Theory and Applications (Y. J. Cho, J. K. Kim and S. S. Dragomir, eds.), vol. 1, Nova Science Publishers Inc., New York, 2002, p. 193–208.

In: Inequality Theory and Applications, Volume 4
Editors: Yeol Je Cho et al., pp.137–151

ISBN 1-59454-874-9

SOME BOUNDS FOR THE LOGARITHMIC FUNCTION

FLEMMING TOPSØE[1]

ABSTRACT. Bounds for the logarithmic function are studied. In particular, we establish bounds with rational functions as approximants. The study leads into the fascinating areas of Padé approximations ([2], [6]), continued fractions ([7], [11]) and orthogonal polynomials ([14], [4]) as well as the somewhat frightening jungle of special functions and associated identities ([5], [9]). Originally, the results aimed at establishing certain inequalities for Shannon entropy but are here discussed in their own right (the applications to entropy inequalities will be published elsewhere).

The exposition is informal, a kind of essay, with only occasional indications of proofs. The reader may take it as an invitation to further studies. Enough details are provided to enable the reader to verify all statements. To the expert in the fields pointed to there is little or nothing new.

1. Basic Inequalities

Consider first the truly basic inequalities:

$$1 - \frac{1}{x} \leq \ln x \leq x - 1, \quad x > 0. \tag{1}$$

Here and in all inequalities below it is understood that the inequalities shown are strict, except in easily recognizable cases of equality. One may prefer to write (1) in the form

$$\frac{x}{1+x} \leq \ln(1+x) \leq x, \quad x > -1. \tag{2}$$

The bounds (1) and (2) involve rational functions of type $[1,1]$ (the left-hand-side) and of type $[1,0]$ (the right-hand-side). Here, the *type* refers to the degrees of numerator and denominator in the rational function concerned.

The simplicity, generality and usefulness of (1) and (2) is unbeatable. Nevertheless, there are many other interesting inequalities for the logarithmic function. For instance, if we separate positive and negative values of x, none of the following two inequalities is much more complex:

$$\frac{2x}{2+x} \leq \ln(1+x) \leq \frac{x}{2} \cdot \frac{2+x}{1+x}, \quad 0 \leq x < \infty, \tag{3}$$

$$\frac{2x}{2+x} \geq \ln(1+x) \geq \frac{x}{2} \cdot \frac{2+x}{1+x}, \quad -1 < x \leq 0 \tag{4}$$

Received March 24, 2004.

2000 *Mathematics Subject Classification.* 41A21, 33C45, 11A55.

Key words and phrases. Logarithmic inequalities, Padé approximation, continued fractions, Jacobi polynomials, Legendre polynomials.

This work was supported by the Danish Natural Science Research Council.

[1] Department of Mathematics, University of Copenhagen, Universitetsparken 5, Denmark. E-mail: topsoe@math.ku.dk

–and these inequalities are sharper than (2). Also note that the first inequality of (3) is an improvement, for $x \geq 0$, over the first inequality of (2) with a rational function of the same type – whereas the second inequality of (2) cannot, of course, be improved in a similar way.

The inequalities (3) and (4) may be written as a single double inequality, e.g. by taking absolute values. Another possibility is to define the function λ by

$$\lambda(x) = \frac{\ln(1+x)}{x} \tag{5}$$

(λ represents the slope of the chorde on the function $x \curvearrowright \ln(1+x)$ which connects $(0,0)$ with $(x, \ln(1+x))$). By continuity, $\lambda(0) = 1$. Then

$$\frac{2}{2+x} \leq \lambda(x) \leq \frac{2+x}{2+2x}, \quad x > -1. \tag{6}$$

For some possibilities to sharpen the bounds (3) and (4), see (22) and (23) further on. However, the form chosen is convenient for a special reason. Actually, (3) and (4) are equivalent as follows by writing $\ln(1+x)$ in the form $-\ln(1-\frac{x}{1+x})$. To further exploit this observation, we introduce a notion of *duality* between functions defined on $[0,\infty[$ and functions defined on $]-1,0]$. The *dual* of one such function, say ϕ, is defined by the relation

$$\phi^*(x) = -\phi\left(\frac{-x}{1+x}\right). \tag{7}$$

Inequalities of the form $\phi(x) \leq \ln(1+x) \leq \psi(x)$ for $x \in [0,\infty[$ (respectively, for $x \in]-1,0]$) then translate into inequalities $\phi^*(x) \geq \ln(1+x) \geq \psi^*(x)$ for $x \in]-1,0]$ (respectively, for $x \in [0,\infty[$).

In (3) and (4) we have an instance with two *self-dual* functions in the sense that for the two analytic functions – say ϕ and ψ – given on the full interval $]-1,\infty[$ by, respectively the left-hand and the right-hand expression in (3) and (4), we find that

$$(\phi_{|[0,\infty[})^* = \phi_{|]-1,0]} \tag{8}$$

and similarly for ψ (here, the subscripts refer to restrictions of the domain of definition). As the functions we shall consider will be analytic in $]-1,\infty[$, we do not find it necessary to indicate the restrictions to the appropriate interval since the analytic form of the functions will be the same whether we consider the restriction to $[0,\infty[$ or to $]-1,0]$. For the functions above we may thus write simply $\phi^* = \phi$ and $\psi^* = \psi$.

Let us return to (2). Also there, duality is relevant. Indeed, writing (2) in the form $\phi(x) \leq \ln(1+x) \leq \psi(x)$, we find that $\phi^* = \psi$ (and $\psi^* = \phi$), i.e. (ϕ, ψ) is a *dual pair*. This explains why there is no restriction on x in (2), except the natural one, $x > -1$.

Duality takes a different – and somewhat simpler – form when having bounds for $\ln x$ in mind rather than bounds for $\ln(1+x)$. In order to save on notation, we use only one symbol, a "star", for duality and then indicate by an overline "tilde" if the functions are intended as bounds for $\ln(x)$ rather than for $\ln(1+x)$. The duality definition for "tilde-functions" is then given by

$$\tilde{\phi}^*(x) = -\tilde{\phi}(x^{-1}). \tag{9}$$

As an example of a basic inequality of $\ln x$-type, we mention

$$|\ln x| \leq \frac{1}{2}\left|x - \frac{1}{x}\right|, \quad x > 0 \tag{10}$$

which is obtained from the right-hand inequalities of (3) and (4).

From (10), and after a further substitution $x := x^a$ with a a positive parameter, we obtain useful approximations of $\ln x$. In more detail, one finds that

$$\frac{1}{2a}(x^a - x^{-a}) \downarrow \ln x \quad \text{as } a \downarrow 0, \quad 1 \le x < \infty, \tag{11}$$

and that

$$\frac{1}{2a}(x^a - x^{-a}) \uparrow \ln x \quad \text{as } a \downarrow 0, \quad 0 < x \le 1. \tag{12}$$

These results – clearly duals of each other – can also be derived directly from the expansion

$$\frac{1}{2a}(x^a - x^{-a}) = \ln x + \sum_{n=1}^{\infty} \frac{a^{2n}}{(2n+1)!}(\ln x)^{2n+1}. \tag{13}$$

Consider, for instance, the case $a = \frac{1}{2}$. This leads to the inequality

$$\ln x \le \frac{x-1}{\sqrt{x}}, \quad 1 \le x < \infty, \tag{14}$$

with reversal of the inequality sign for $0 < x < 1$. Expressed in terms of the λ-function this shows that

$$\lambda(x) \le \frac{1}{\sqrt{1+x}}, \quad x > -1. \tag{15}$$

This useful inequality – which one could just as well had been led to by considering the product of the extreme terms in (6) – goes back at least to Karamata [8], cf. also Mitrinović [12, Section 3.6.15].

Let us end this section with a more special inequality which is, again, related to (6). It is our only result which involves two parameters (other such inequalities can be found in [12]). The inequality states that for $0 \le x \le 1$ and $0 \le y < \infty$,

$$(2-x)\lambda(y) - \frac{1-x}{1+y} \le \lambda(xy) \le x\lambda(y) + (1-x). \tag{16}$$

This follows by a standard analysis of the inequalities which result when you keep y fixed (for the proof, consider the function equal to the difference between terms of the inequality you wish to prove, and differentiate twice). The indicated proof makes use of (6) (which is needed in order to discuss endpoint behaviour). One may also note that, for $0 \le x \le 1$, the second inequality in (6) follows by comparison of the two extreme terms in (16).

2. Rational Approximants for $\ln(1+x)$

We have demonstrated the usefulness of (6) and now turn to a more systematic study of similar bounds with more general rational functions. We choose to focus on bounds for $\ln(1+x)$ for $x \ge 0$. We already have some examples of lower and upper bounds of $\ln(1+x)$ and realize that it is natural to seek lower bounds of type $[n, n]$ and upper bounds of type $[n, n-1]$. We shall find such bounds for each $n \ge 1$.

To find criteria for the selection of good lower bounds, assume that $\phi(x) = \frac{F}{G}$ is such a bound with F and G polynomials, both of degree n, say. We insist that the bound is exact for

$x = 0$, hence the constant term in F is 0. For the difference function $\delta(x) = \ln(1+x) - \phi(x)$, we find that

$$\delta'(x) = \frac{G(x)^2 - (1+x)\big(F'(x)G(x) - F(x)G'(x)\big)}{(1+x)G(x)^2}. \tag{17}$$

For a loose consideration, let us neglect the fact that the denominator here varies with the choice of ϕ. Then we only need to assure that the numerator is small and non-negative. To choose among the possible bounds, note that the numerator is a polynomium of degree $2n$. We aim at a bound which is especially good for small values of x and realize that we should attempt to make lower terms vanish. We achieve this by insisting that *all* terms in the numerator vanish, except the leading term.

To be precise, we denote by ϕ_n the function of the form $\phi_n = \frac{F}{G}$ with F and G two polynomials of degree n such that $F(0) = 0$ and $G(0) > 0$ and such that the numerator in (17) is a positive constant times x^{2n}. Similarly, ψ_n is defined by $\psi_n = \frac{F}{G}$, where F is a polynomial of degree n with $F(0) = 0$ and G a polynomial of degree $n-1$ with $G(0) > 0$ and such that the numerator in (17) is a negative constant times x^{2n-1}.

It turns out that, for each $n \geq 1$, ϕ_n and ψ_n are uniquely determined by these requirements. We introduce the following standard representations:

$$\phi_n(x) = \frac{xP_{n-1}(x)}{Q_n(x)}, \quad \psi_n(x) = \frac{xR_{n-1}(x)}{S_{n-1}(x)} \tag{18}$$

with $(P_n)_{n\geq 0}$, $(Q_n)_{n\geq 0}$, $(R_n)_{n\geq 0}$ and $(S_n)_{n\geq 0}$ four sets of polynomials of degrees as indicated by the subscript and such that

$$Q_{n,n} = 1\,, \quad R_{n-1,n-1} = 1 \tag{19}$$

(with natural notation for the coefficients of the polynomials we consider). For convenience, we put $Q_0(x) \equiv 1$. By (19), the leading term in Q_n is x^n and the leading term in R_{n-1} is x^{n-1}, i.e., Q_n and R_{n-1} are *monic* polynomials. Observe that we normalize via the denominator in ϕ_n and via the numerator in ψ_n. With the chosen normalization of the representations for the ϕ- and ψ-functions, the P-, Q-, R- and S-polynomials are all uniquely determined. To save on notation, we shall normally write P_n (respectively, Q_n, R_n, S_n and also ϕ_n and ψ_n) instead of $P_n(x)$ (respectively, $Q_n(x), R_n(x)$ etc.). We often refer to the four sets of polynomials as the *PQRS-polynomials.*

To recapitulate, ϕ_n and ψ_n are defined by (18) and (19) and by the essential requirements

$$Q_n^2 - (1+x)\big((xP_{n-1})'Q_n - xP_{n-1}Q_n'\big) = x^{2n}, \tag{20}$$

$$(1+x)\big((xR_{n-1})'S_{n-1} - xR_{n-1}S_{n-1}'\big) - S_{n-1}^2 = S_{n-1,n-1}x^{2n-1}. \tag{21}$$

Clearly, ϕ_n is a lower- and ψ_n an upper bound of $\ln(1+x)$ for $x \in [0,\infty[$. This follows as the functions $x \curvearrowright \ln(1+x) - \phi_n(x)$ and $x \curvearrowright \psi_n(x) - \ln(1+x)$ both vanish at $x = 0$ and have positive derivatives in $]0,\infty[$.

The considerations leading to the above definitions express a key idea in the theory of *Padé approximation*, cf. Baker and Graves-Morris [2], a standard reference. In the terminology of that theory, ϕ_n is the $[n,n]$-*Padé approximant* and ψ_n the $[n,n-1]$-*Padé approximant* of $\ln(1+x)$. We shall refer to the ϕ_n's as simply the *lower approximants* (to $\ln(1+x)$) and to the ψ_n's as the *upper approximants.*

Instead of just referring the reader to the literature on Padé approximation, we base our exposition on experiments facilitated by modern computing tools. This will lead rather quickly to desired formulas and other insights. Full proofs of relations found experimentally may not always be so obvious. We shall include enough details to enable the reader to validate all statements in the usual rigorous mathematical style. The interested reader will find further results in the literature referred to.

3. Some Experiments

In order to get a feel for the nature of the approximants defined in the previous section, it is natural to work out a number of examples. This can be done by equating coefficients of polynomials occuring in the defining relations (20) and (21). With hinsight, this can be done much more conveniently by recursion formulas developed later ((60)–(63)) or by simply asking MAPLE to work out the relevant Padé approximants. Anyhow, in one way or another we can get at the first few approximants and obtain a suitable table, cf. Table 1.

For instance, $\phi_1(x) \leq \ln(1+x)$ for $x \geq 0$ which we recognize as the first inequality of (3). As $\phi_1^* = \phi_1$, the first inequality of (4) also follows. The determination of ψ_2 gives the inequality

$$\ln(1+x) \leq \frac{x(6+x)}{2(3+2x)}, \quad x \geq 0, \tag{22}$$

a strengthening of the second inequality in (3) with a bound of the same type ($[2,1]$). And, if we dualize the inequality for ψ_2, we get $\ln(1+x) \geq \psi_2^*(x)$ for $-1 < x \leq 0$, i.e.

$$\ln(1+x) \geq \frac{x(6+5x)}{2(1+x)(3+x)}, \quad -1 < x \leq 0, \tag{23}$$

a streghtening of the second inequality of (4), though with a function of a different type ($[2,2]$ rather than $[2,1]$).

By (18) and (19), Table 1 reveals the identity of the first $PQRS$-polynomials, cf. Table 2.

TABLE 1: Lower- and upper approximants to $\ln(1+x)$

n	ϕ_n	ψ_n
1	$\frac{2x}{2+x}$	x
2	$\frac{3x(2+x)}{6+6x+x^2}$	$\frac{x(6+x)}{2(3+2x)}$
3	$\frac{x(60+60x+11x^2)}{3(20+30x+12x^2+x^3)}$	$\frac{x(30+21x+x^2)}{3(10+12x+3x^2)}$
4	$\frac{5x(84+126x+52x^2+5x^3)}{6(70+140x+90x^2+20x^3+x^4)}$	$\frac{x(420+510x+140x^2+3x^3)}{12(35+60x+30x^2+4x^3)}$
5	$\frac{x(7560+15120x+9870x^2+2310x^3+137x^4)}{30(252+630x+560x^2+210x^3+30x^4+x^5)}$	$\frac{x(3780+6510x+3360x^2+505x^3+6x^4)}{30(126+280x+210x^2+60x^3+5x^4)}$
6	$\frac{7x(1320+3300x+2960x^2+\ldots}{10(924+2772x+3150x^2+\ldots}$ $\frac{\ldots+1140x^3+174x^4+7x^5)}{\ldots+1680x^3+420x^4+42x^5+x^6)}$	$\frac{x(13860+30870x+23520x^2+7035x^3+672x^4+5x^5)}{30(462+1260x+1260x^2+560x^3+105x^4+6x^5)}$

TABLE 2: $PQRS$-polynomials of degrees 0, 1, 2 and 3.

n	0	1	2	3
P_n	2	$6+3x$	$20+20x+\frac{11}{3}x^2$	$70+105x+\frac{130}{3}x^2+\frac{25}{6}x^3$
Q_n	1	$2+x$	$6+6x+x^2$	$20+30x+12x^2+x^3$
R_n	1	$6+x$	$30+21x+x^2$	$140+170x+\frac{140}{3}x^2+x^3$
S_n	1	$6+4x$	$30+36x+9x^2$	$140+240x+120x^2+16x^3$

A graphical plot of the first four approximants is shown in Figure 1. After more extensive plotting – if the reader is not yet convinced – one will realize that the ϕ_n's increase with n and the ψ_n's decrease, both sequences with $\ln(1+x)$ as limit function. In order to test this, we look at differences of the relevant functions. One finds that

$$\psi_n - \phi_n = \frac{x^{2n}}{Q_n S_{n-1}}, \tag{24}$$

$$\psi_{n+1} - \phi_n = \frac{x^{2n+1}}{Q_n S_n}, \tag{25}$$

$$\phi_{n+1} - \phi_n = \frac{2}{n+1}\frac{x^{2n+1}}{Q_{n+1} Q_n}, \tag{26}$$

$$\psi_n - \psi_{n+1} = (2n+1)\frac{x^{2n}}{S_{n-1} S_n}. \tag{27}$$

This may be obtained experimentally but also follows from the definitions of the approximants (except for the constants in (26) and (27) which depend on the later relations (29) and (30)).

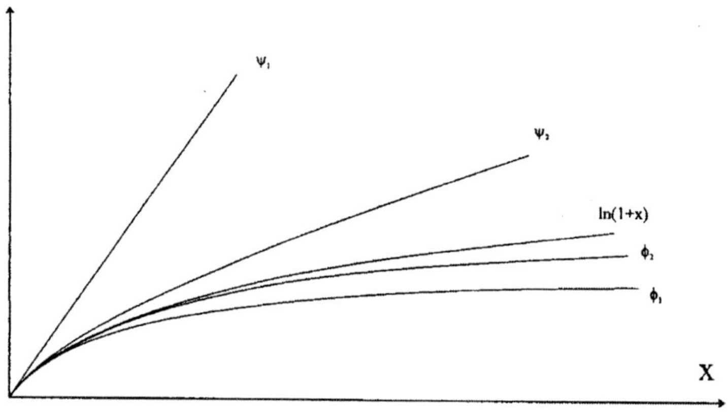

FIGURE 1. Upper and lower bounds of $\ln(1+x)$ for $x \geq 0$.

In order to find good approximations of $\ln(1+x)$ (for $x \geq 0$), we could proceed differently from the above approach which is tied to the defining relations (20) and (21). As a start, take x as a reasonable approximation of $\ln(1+x)$. This is too much, thus in the next step we divide by 1 plus some non-negative function. Not to increase the complexity too much, take this function to be proportional to x, say cx. Then we get a type-$[1,1]$ approximation of $\ln(1+x)$. If this is set to ϕ_1 – known to be a reasonable bound – we get $c = \frac{1}{2}$. Now, ϕ_1 is too small, hence in our next step we divide x by 2 plus a non-negative function. Choosing again a function proportional to x, we obtain a type-$[2,1]$ bound and it is reasonable to adjust this so that it coincides with ψ_2. We now have the bound

$$\psi_2(x) = \frac{x}{1 + \frac{x}{2+\frac{x}{3}}}$$

and may continue based on the identification of the first few good bounds given in Table 1.

The approach just described is of course well known and nothing but an attempt to represent $\ln(1+x)$ by a *continued fraction.* The approach above does not lead in a unique way to (the beginnings of) a continued fraction. E.g., above we could choose to look, as we did, at x divided by 2 + something or we could have looked at $\frac{1}{2}x$ divided by 1 plus something. Experimenting with the possibilities, you soon see that a very simple structure emerges if you at each step divide by n plus something. Indeed, you then arrive at the beautiful representation

$$\ln(1+x) = \frac{x}{1} + \frac{1^2 x}{2} + \frac{1^2 x}{3} + \frac{2^2 x}{4} + \frac{2^2 x}{5} + \cdots + \frac{n^2 x}{2n} + \frac{n^2 x}{2n+1} + \cdots . \tag{28}$$

This representation is not new. You find it in [3], [4], [2], [7] and [11]. In fact, the result is more than two hundred years old (!) and goes back to Lambert, cf. [10]. Regarding historical comments – to this and other aspects of the paper – we refer to [3] and to [7].

In a sense, the continued fraction (28) tells us all we need to know as it provides us with easy access to the approximants ϕ_n and ψ_n. We shall elaborate more on this in Section 5. For now, let us look at a few more "experiments".

Some coefficients for the $PQRS$-polynomials are easy to guess from Tables 1 and 2, e.g., we see that

$$S_{n,n} = (n+1)^2 . \tag{29}$$

As another example, we realize that $P_{n,n} - P_{n-1,n-1} = \frac{2}{n+1}$, hence

$$P_{n,n} = 2 \sum_{\nu=1}^{n+1} \frac{1}{\nu} . \tag{30}$$

One may also go hunting for relationships among the coefficients by consulting the "On-Line Encyclopedia of Integer Sequences", cf. Sloane [13]. For instance, enquiring about the sequence $6, 30, 90, 210$, the feed-back from this source will reveal the fact that

$$Q_{n,n-2} = \frac{1}{4}[n+2]_4 .$$ [1]

Many other relations for the $PQRS$-coefficients can be discovered in this way. Instead of pursuing this line of investigation, we shall have a look at the approximants after a transformation so that they approximate $\ln x$ rather than $\ln(1+x)$. In other words, we are asking about the "tilde-functions"

$$\tilde{\phi}_n(x) = \phi_n(x-1); \quad \tilde{\psi}_n(x) = \psi_n(x-1). \tag{31}$$

TABLE 3: Lower- and upper approximants to $\ln(x)$

n	$\tilde{\phi}_n$	$\tilde{\psi}_n$
1	$\frac{2(x-1)}{1+x}$	$x-1$
2	$\frac{3(x-1)(1+x)}{1+4x+x^2}$	$\frac{(x-1)(x+5)}{2(1+2x)}$
3	$\frac{(x-1)(11+38x+11x^2)}{3(1+9x+9x^2+x^3)}$	$\frac{(x-1)(10+19x+x^2)}{3(1+6x+3x^2)}$
4	$\frac{5(x-1)(5+37x+37x^2+5x^3)}{6(1+16x+36x^2+16x^3+x^4)}$	$\frac{(x-1)(47+239x+131x^2+3x^3)}{12(1+12x+18x^2+4x^3)}$
5	$\frac{(x-1)(137+1762x+3762x^2+1762x^3+137x^4)}{30(1+25x+100x^2+100x^3+25x^4+x^5)}$	$\frac{(x-1)(131+1281x+1881x^2+481x^3+6x^4}{30(1+20x+60x^2+40x^3+5x^4)}$
6	$\frac{7(x-1)(7+139x+514x^2+\dots}{10(1+36x+225x^2+400x^3+\dots}$	$\frac{(x-1)(142+2272x+6397x^2+\dots}{30(1+30x+150x^2+200x^3+\dots}$
	$\frac{\dots+514x^3+139x^4+7x^5)}{\dots+225x^4+36x^5+x^6)}$	$\frac{\dots+4397x^3+647x^4+5x^5)}{\dots+75x^4+6x^5)}$

[1]The notation $[a]_k$ is used for the *descending factorial* $a(a-1)\cdots(a-k+1)$.

This leads to the functions in Table 3.

TABLE 4: Zeroes of the $PQRS$-polynomials.

P_1	-2.0000				
P_2	-4.1356	-1.3189			
P_3	-7.2397	-2.0000	-1.1603		
P_4	-11.3204	-2.9243	-1.5197	-1.0969	
P_5	-16.3907	-4.0764	-2.0000	-1.3251	-1.0650
Q_1	-2.0000				
Q_2	-4.7321	-1.2679			
Q_3	-8.8730	-2.0000	-1.1270		
Q_4	-14.4026	-3.0302	-1.4926	-1.0746	
Q_5	-21.3174	-4.3334	-2.0000	-1.3000	-1.0492
R_1	-6.0000				
R_2	-19.4582	-1.5418			
R_3	-42.7683	-2.6743	-1.2240		
R_4	-77.0830	-4.2481	-1.7114	-1.1242	
R_5	-123.2945	-6.2526	-2.3648	-1.4089	-1.0792
S_1	-1.5000				
S_2	-2.8165	-1.1835			
S_3	-4.7094	-1.6934	-1.0992		
S_4	-7.1551	-2.4015	-1.3828	-1.0606	
S_5	-10.1487	-3.2837	-1.7793	-1.2469	-1.0415

We end this section by an investigation of the zeroes of the $PQRS$-polynomials. The result – found by numerical computation using MAPLE – is reported in Table 4. We realize that the *interlacing property* holds for each set of polynomials, e.g., between adjacent zeroes of P_n you find a zero of P_{n-1}. And also some "mixed interlacing" takes place, as we shall return to at the very end of the paper.

4. Some Facts

In this section it is assumed that any occurring x is non-negative unless stated otherwise explicitly.

The experiments of the previous section lead to important facts about the ϕ_n's and ψ_n's. Let us start by looking at the quality of these functions as bounds for $\ln(1+x)$.

Apparently, all coefficients in the *PQRS*-polynomials are positive. From (24)-(27), it therefore follows that

$$\phi_1(x) \le \phi_2(x) \le \cdots \le \ln(1+x) \le \cdots \le \psi_2(x) \le \psi_1(x). \tag{32}$$

Furthermore, by (25), recalling also (19) and (29), we see that

$$\psi_{n+1}(x) - \phi_n(x) \le \frac{x}{(n+1)^2}. \tag{33}$$

Though this bound is quite loose, it is sharp enough to imply that, for each $x \ge 0$, the sequences $\phi_n(x)$ and $\psi_n(x)$ both converge to $\ln(1+x)$ as $n \to \infty$. We also note that

$$\ln(1+x) - \phi_n(x) \le \frac{x^{2n+1}}{Q_n(x)S_n(x)}, \tag{34}$$

$$\psi_n(x) - \ln(1+x) \le \frac{x^{2n}}{Q_n(x)S_{n-1}(x)}. \tag{35}$$

These bounds are less explicit but typically much sharper than (33).

The behaviour of $\phi_n(x)$ and $\psi_n(x)$ for large x is given by the relations

$$\phi_n(x) \approx 2\sum_{\nu=1}^{n} \frac{1}{\nu}; \quad \psi_n(x) \approx \frac{x}{n^2} \tag{36}$$

in the sense that the ratios involved converge to 1 for $x \to \infty$. These relations follow from (29) and (30).

We also note that the functions ϕ_n and ψ_n are increasing. As for ϕ_n this follows from (20) (see also (17)) which tells us that

$$\phi_n'(x) = \frac{Q_n(x)^2 - x^{2n}}{(1+x)Q_n(x)^2} \ge 0. \tag{37}$$

The approximants may be characterized in a natural way which is quite different from the definitions via (20) and (21). Indeed, for each $n \ge 1$, ϕ_n is the unique type-$[n,n]$ rational function with $\phi_n(0) = 0$ and $\phi_n(x) \le \ln(1+x)$ for $x \ge 0$ which dominates any other such function *locally*, i.e., for any such function f, there exists $\varepsilon > 0$ such that $\phi_n(x) \ge f(x)$ for $0 \le x < \varepsilon$. In fact, for this conclusion we may assume about type-$[n,n]$ functions f considered only that $f \le \psi_{n+1}$, locally, instead of $f(x) \le \ln(1+x)$ for $x \ge 0$. A similar characterization holds for ψ_n.

Note that we have to work locally with inequalities in the characterizations just pointed out (consider, for example, the function given by $f(x) = \frac{3x}{4+x}$; then $f(x) \le \ln(1+x)$ for $x \ge 0$ – as $f \le \phi_2$ – but $f \le \phi_1$ does not hold globally for $x \ge 0$, only locally). The proofs of the characterizations rest on the facts that if a type-$[n,n]$-function is not identical to ϕ_n, then there is a $\nu \le 2n$ such that $f^{(\nu)}(0) \ne \phi_n^{(\nu)}(0)$ and if a type-$[n,n-1]$ function g is not identical to ψ_n, then there is a $\nu \le 2n-1$ such that $g^{(\nu)}(0) \ne \psi_n^{(\nu)}(0)$.

The determination of the *PQRS*-polynomials in closed form is not that obvious based on our experiments. Most strikingly, perhaps, is that if we define $\tilde{Q}_n$ by

$$\tilde{Q}_n(x) = Q_n(x-1),$$

then

$$\tilde{Q}_n(x) = \sum_{k=0}^{n} \binom{n}{k}^2 x^k$$

(cf. Table 3). As $Q_n(x) = \tilde{Q}_n(1+x)$, this leads to the formula

$$Q_{n,k} = \binom{n}{k} \sum_{\nu=0}^{n-k} \binom{n-k}{\nu} \binom{n}{n-k-\nu}.$$

The sum here can be evaluated by using the *binomial Chu-Vandermonde formula*:

$$\sum_{j=0}^{N} \binom{N}{j} [x]_j [y]_{N-j} = [x+y]_N$$

or, more directly, by the equivalent *Chu-Vandermonde convolution* formula:

$$\sum_{j=0}^{N} \binom{x}{j} \binom{y}{N-j} = \binom{x+y}{N}.$$

For these identities, see [1]. What we find is that

$$Q_{n,k} = \binom{n}{k} \binom{2n-k}{n-k} = \binom{2n-k}{k, n-k, n-k}. \tag{38}$$

Thus the coefficients in Q_n have been identified as certain trinomial coefficients.

Having determined the $Q_{n,k}$'s we may use the defining relation (20) to determine the $P_{n-1,k}$'s. With some effort this leads to the formula

$$P_{n-1,k} = \sum_{\nu=0}^{k} \frac{(-1)^\nu}{\nu+1} Q_{n,k-\nu}. \tag{39}$$

Combined with (38) this gives a reasonable determination of the coefficients of the P-polynomials, e.g. it shows that $P_{n-1,0} = Q_{n,0} = \binom{2n}{n}$. However, the formulas can sometimes be simplified (consider the case $k = n-1$ when $P_{n-1,n-1} = 2\sum_1^n \frac{1}{\nu}$ according to (30)).

A striking feature of the $\tilde{phi}_n$-functions is that they are self-dual. This is seen from Table 3. Indeed, writing

$$\tilde{phi}_n(x) = \frac{(x-1)\tilde{P}_{n-1}(x)}{\tilde{Q}_n(x)}$$

with $\tilde{P}_{n-1}(x) = P_{n-1}(x-1)$ and $\tilde{Q}_n(x) = Q_n(x-1)$, we realize that these polynomials are *self-reflected* in the sense that

$$x^{n-1}\tilde{P}_{n-1}(\frac{1}{x}) = \tilde{P}_{n-1}(x),$$

$$x^n \tilde{Q}_n(\frac{1}{x}) = \tilde{Q}_n(x).$$

By simple calculation, this implies that the $\tilde{phi}_n$'s are self-dual. Hence also the ϕ_n's are self-dual: $\phi_n^* = \phi_n$. By duality, this implies that ϕ_n is a best type-$[n,n]$ upper bound of $\ln(1+x)$ for $-1 < x \le 0$ (with "best" being understood in much the same way as discussed for the bounds found for $x \ge 0$).

The functions ψ_n are not self-dual. The duals ψ_n^* are even of a different type, viz. of type $[n,n]$. These functions then, are the best type-$[n,n]$ lower bounds of $\ln(1+x)$ for $-1 < x \le 0$.

5. Exploiting Lamberts Continued Fraction

It is natural to seek procedures which allow recursive determination of the *PQRS*-polynomials. This can be achieved based on the defining relations and the derived relations (24)–(27). However, this will, typically, involve products of polynomials and simpler procedures are desirable. Here Lamberts expansion (28) provides the proper tool.

Following usual terminology of continued fraction theory, let A_n and B_n denote the *approximants* related to (28). These are polynomials associated in the natural manner to the corresponding finite continued fractions, e.g.

$$\frac{A_1}{B_1} = \frac{x}{1}\,, \quad \frac{A_2}{B_2} = \frac{x}{1+\frac{1^2x}{2}}\,, \quad \frac{A_3}{B_3} = \frac{x}{1+\frac{1^2x}{2+\frac{1^2x}{3}}}\,.$$

The first polynomials are

$$A_0 = 0\,,\ A_1 = x\,,\ B_0 = 1\,,\ B_1 = 1 \tag{40}$$

and, exploiting the special form of (28), the following recursive relations hold for $n \ge 1$:

$$A_{2n} = 2nA_{2n-1} + n^2xA_{2n-2}\,, \tag{41}$$

$$A_{2n+1} = (2n+1)A_{2n} + n^2xA_{2n-1}\,, \tag{42}$$

$$B_{2n} = 2nB_{2n-1} + n^2xB_{2n-2}\,, \tag{43}$$

$$B_{2n+1} = (2n+1)B_{2n} + n^2xB_{2n-1}\,. \tag{44}$$

We realize that, for $n \ge 1$,

$$\phi_n = \frac{A_{2n}}{B_{2n}}\,, \quad \psi_n = \frac{A_{2n-1}}{B_{2n-1}}\,. \tag{45}$$

The polynomials A_{2n-1}, A_{2n}, B_{2n}, and B_{2n+1} are all of degree n and for basic coefficients we find

$$A_{n,0} = 0\,, \quad A_{n,1} = B_{n,0} = n!\,, \tag{46}$$

$$A_{2n+1,n+1} = B_{2n,n} = (n!)^2\,. \tag{47}$$

Together with (45), (47) tells us that

$$xP_{n-1} = \frac{A_{2n}}{(n!)^2}\,,\ Q_n = \frac{B_{2n}}{(n!)^2}\,, \tag{48}$$

$$xR_{n-1} = \frac{A_{2n-1}}{((n-1)!)^2}\,,\ S_{n-1} = \frac{B_{2n-1}}{((n-1)!)^2}\,. \tag{49}$$

Now, Q_n is known, cf. (38), hence also B_{2n} can be found. And then, from (43), we can determine B_{2n-1}, hence, by (49), also S_{n-1}. This leads to the formula

$$S_{n-1} = n\sum_{\nu=0}^{n-1}\binom{2n-\nu-1}{\nu,\,n-\nu,\,n-\nu-1}x^\nu\,. \tag{50}$$

Similarly, a formula for R_{n-1} can be found based on (39), (41) and (49). Going through the details you find that

$$R_{n-1} \\ = n\sum_{\nu=0}^{n-1}\sum_{\mu=1}^{\nu+1}(-1)^{\mu-1}\binom{2n-\nu+\mu-2}{\nu-\mu+1, n-\nu+\mu-1, n-\nu+\mu-2}\mu^{-1}x^{\nu}. \tag{51}$$

The expressions (38), (39), (50) and (51) thus provide formulas for the $PQRS$-polynomials in closed form with very satisfactory expressions for the denominator polynomials (Q, S) and somewhat less satisfactory formulas for the numerator polynomials (P, R).

For practical calculations, however, the recursive relations (41)–(44) are more expedient and simple to programme. By (48) and (49) these relations may be written directly in terms of the $PQRS$-polynomials:

$$P_{n-1} = \frac{2}{n}R_{n-1} + xP_{n-2}\,, \tag{52}$$

$$R_n = (2n+1)P_{n-1} + xR_{n-1}\,, \tag{53}$$

$$Q_n = \frac{2}{n}S_{n-1} + xQ_{n-1}\,, \tag{54}$$

$$S_n = (2n+1)Q_n + xS_{n-1}\,. \tag{55}$$

Here, $n \geq 1$ and for (52) we have put $P_{-1} = 0$.

The special structure of these recurrence relations may be emphasized by introducing the notation ΔP_n for the polynomial

$$(\Delta P_n)(x) = P_n(x) - xP_{n-1}(x)$$

and similarly for ΔQ_n etc. Then,

$$\Delta P_{n-1} = \frac{2}{n}R_{n-1}\,; \quad \Delta R_n = (2n+1)P_{n-1}\,, \tag{56}$$

$$\Delta Q_n = \frac{2}{n}S_{n-1}\,; \quad \Delta S_n = (2n+1)Q_n\,. \tag{57}$$

Here $n \geq 1$ (with $\Delta P_0 = 2$).

Another form of (52)–(55) is obtained for the *reflected polynomials*. Following [6], the reflected polynomial of a polynomial A is the polynomial denoted $\bar{A}$ which is given by

$$\bar{A}(x) = x^N A(x^{-1})$$

with N the degree of A. We find:

$$\bar{P}_{n-1} - \bar{P}_{n-2} = \frac{2}{n}\bar{R}_{n-1}\,; \quad \bar{R}_n - \bar{R}_{n-1} = (2n+1)x\bar{P}_{n-1}\,, \tag{58}$$

$$\bar{Q}_n - \bar{Q}_{n-1} = \frac{2}{n}x\bar{S}_{n-1}\,; \quad \bar{S}_n - \bar{S}_{n-1} = (2n+1)\bar{Q}_n\,. \tag{59}$$

6. Relations to Orthogonal Polynomials

The recurrence relations (52)–(55) are ideal for recursive computation of the $PQRS$-polynomials. In (52) and (53), numerator polynomials are "entangled", and in (54) and

(55) so are denominator polynomials. It is, however easy to "unentangle" the formulas so as to only involve one set of polynomials at a time. In this way you find, for $n \geq 2$:

$$(n+1)P_n - (2n+1)(x+2)P_{n-1} + nx^2 P_{n-2} = 0\,, \tag{60}$$

$$nQ_n - (2n-1)(x+2)Q_{n-1} + (n-1)x^2 Q_{n-2} = 0\,, \tag{61}$$

$$(2n-1)R_n - \big(4nx + \frac{2}{n}(4n^2-1)\big)R_{n-1} + (2n+1)x^2 R_{n-2} = 0\,, \tag{62}$$

$$(2n-1)S_n - \big(4nx + \frac{2}{n}(4n^2-1)\big)S_{n-1} + (2n+1)x^2 S_{n-2} = 0\,. \tag{63}$$

In passing we note that it is easy to prove from (60) and (61) that the ϕ_n's are self-dual.

It is well known that orthogonal polynomials satisfy three-term recurrence relations. With the above formulas in mind it therefore lies nearby to investigate the possible relations of the *PQRS*-polynomials to orthogonal polynomials. However, the recurrence relations (60)-(63) are not in standard form as known from the theory of orthogonal polynomials because of the appearence of x in the last term. This points to the desirability of rewriting the formulas by introducing a suitable transformation.

Such a transformation may be suggested by looking at the zeroes of the *PQRS*-polynomials, recalling that these are dense in the support of any underlying measure. Here, the information provided by Table 4 is useful. It shows that the zeroes lie in $]-\infty,-1[$ and also, that the "interlacing" behaviour which is characteristic for orthogonal polynomials applies to our polynomials.

Possibly after some experimentation, it is found that a transformation which places the zeroes in the interval $]-1,1[$ have a simplifying effect. This may be achieved by first making a transformation to the reflected polynomials, which brings the zeroes into $]-1,0[$, and then consider the natural linear transformation into $]-1,1[$. We therefore define new polynomials, indicated by the letters $\mathbb{P}, \mathbb{Q}, \mathbb{R}, \mathbb{S}$. For the $\mathbb{P}$-polynomials, definition and notation is as follows:

$$\mathbb{P}_n(x) = \bar{P}_n\Big(\frac{1}{2}(x-1)\Big) = 2^{-n}(x-1)^n P_n\Big(\frac{2}{x-1}\Big)\,.$$

The other polynomials are defined similarly. We find:

$$\mathbb{P}_n - \mathbb{P}_{n-1} = \frac{2}{n+1}\mathbb{R}_n\,, \tag{64}$$

$$\mathbb{Q}_n - \mathbb{Q}_{n-1} = \frac{x-1}{n}\mathbb{S}_{n-1}\,, \tag{65}$$

$$\mathbb{R}_n - \mathbb{R}_{n-1} = (n+\frac{1}{2})(x-1)\mathbb{P}_{n-1}\,, \tag{66}$$

$$\mathbb{S}_n - \mathbb{S}_{n-1} = (2n+1)\mathbb{Q}_n\,, \tag{67}$$

and the three-term recurrence relations become

$$(n+1)\mathbb{P}_n - (2n+1)x\mathbb{P}_{n-1} + n\mathbb{P}_{n-2} = 0\,, \tag{68}$$

$$n\mathbb{Q}_n - (2n-1)x\mathbb{Q}_{n-1} + (n-1)\mathbb{Q}_{n-2} = 0\,, \tag{69}$$

$$n(2n-1)\mathbb{R}_n - ((4n^2-1)x+1)\mathbb{R}_{n-1} + n(2n+1)\mathbb{R}_{n-2} = 0\,, \tag{70}$$

$$n(2n-1)\mathbb{S}_n - ((4n^2-1)x+1)\mathbb{S}_{n-1} + n(2n+1)\mathbb{S}_{n-2} = 0\,. \tag{71}$$

Note the disappearence of any x's in the last terms. We may apply (68)-(71) with $n \geq 2$ in connection with the start polynomials given in Table 5.

TABLE 5: $\mathbb{PQRS}$-polynomials of degrees 0 and 1.

n	$\mathbb{P}_n$	$\mathbb{Q}_n$	$\mathbb{R}_n$	$\mathbb{S}_n$
0	2	1	1	1
1	$3x$	x	$3x-2$	$3x+1$

We shall relate the $\mathbb{PQRS}$-polynomials to the classical *Jacobi polynomials*, here denoted $P_n^{\alpha,\beta}$, which are associated with the measures on $[-1,1]$ with densities $(1-x)^\alpha(1+x)^\beta$. From standard sources, e.g. [14], we see that these polynomials may be determined from the recurrence relations

$$P_n^{\alpha,\beta} - \frac{(2n-1+\alpha+\beta)\big((2n+\alpha+\beta)(2n+\alpha+\beta-2)x+(\alpha^2-\beta^2)\big)}{2n(n+\alpha+\beta)(2n-2+\alpha+\beta)}P_{n-1}^{\alpha,\beta} \tag{72}$$
$$+\frac{(n+\alpha-1)(n+\beta-1)(2n+\alpha+\beta)}{n(n+\alpha+\beta)(2n-2+\alpha+\beta)}P_{n-2}^{\alpha,\beta} = 0$$

for $n \geq 2$ in conjunction with the first polynomials which are given by

$$P_0^{\alpha,\beta} = 1\,; \quad P_1^{\alpha,\beta} = \frac{1}{2}\big((\alpha+\beta+2)x+(\alpha-\beta)\big)\,. \tag{73}$$

The Jacobi polynomials have a simple expression in closed form:

$$P_n^{\alpha,\beta} = \frac{1}{2^n}\sum_{\nu=0}^{n}\binom{n+\alpha}{\nu}\binom{n+\beta}{n-\nu}(x-1)^{n-\nu}(x+1)^\nu. \tag{74}$$

We realize that

$$\mathbb{Q}_n = P_n^{0,0}\,, \tag{75}$$

i.e., the $\mathbb{Q}_n$'s are nothing but the classical *Legendre polynomials.* By (65), the $\mathbb{S}$-polynomials are closely related to these polynomials. However, apart from a constant factor, they may also be identified directly as orthogonal polynomials. Indeed, by Table 5, (71), (72) and (73) it is easy to check that

$$\mathbb{S}_n = (n+1)P_n^{1,0}\,. \tag{76}$$

The $\mathbb{P}$- and $\mathbb{R}$-polynomials are harder to identify. We appeal to [4] for a systematic table of orthogonal polynomials. The table works with monic polynomials. Let the monic polynomials corresponding to the $\mathbb{P}_n$'s be denoted $[\mathbb{P}_n]$, i.e.

$$[\mathbb{P}_n] = \frac{2^{n-1}}{\binom{2n+1}{n}}\mathbb{P}_n\,. \tag{77}$$

Then you find the recursion formula

$$[\mathbb{P}_n] - x[\mathbb{P}_{n-1}] + \frac{n^2}{4n^2-1}[\mathbb{P}_{n-2}] = 0\,. \tag{78}$$

By a look-up in [4] (case 5, p, 219), you realize that the $[\mathbb{P}_n]$'s are orthogonal polynomials associated with the somewhat bizarre measure on $]-1,1[$ with density function

$$\frac{4}{\pi^2 + \left(\ln \frac{1+x}{1-x}\right)^2}.$$

We have not identified the $\mathbb{R}$-polynomials in a similar manner. Possibly, it is more reasonable to look at the polynomials $\mathbb{S}_n - \mathbb{R}_n$ when searching for properties related to orthogonal polynomials.

Lastly we note two interlacing properties with mixed polynomials. For this, we may as well return to the original $PQRS$-polynomials. In fact, from the defining relation (20) and an investigation of the sign of P_{n-1} at zeroes of Q_n it follows that between any two neighbouring zeroes of Q_n there is a zero of P_{n-1} (and of course only one zero). And from (21) a similar investigation shows that between any two neighbouring zeroes of S_{n-1} there is a zero of R_{n-1}. This accounts for $n-2$ zeroes of R_{n-1}. As is easily seen, the last zero of R_{n-1} is smaller than the smallest zero of S_{n-1}.

Acknowledgement

Thanks are due to Jacob Stordal Christiansen for some of the references to the literature and for identification of the associated orthogonal polynomials related to the P_n's.

References

1. E. S. Andersen and M. E. Larsen, *Combinatorial Summation Identities*, In Analysis, algebra, and computers in mathematical research, volume 156 of Lecture Notes in Pure and Applied Mathematics, Marcel Dekker, 1994, pp. 1–23.
2. G. A. Baker and P. Graves-Morris, *Padé Approximants, volume 59 of Encyclopedia of Mathematics and its Applications, 2nd edition*, Cambridge University Press, 1996.
3. C. Brezinski, *History of Continued Fractions and Padé Approximants*, Springer, Berlin Heidelberg, 1978.
4. T. S. Chihara, *An Introduction to Orthogonal Polynomials*, Gordon and Breach, New York London Paris, 1978.
5. R. Askey, G. E. Andrews and R. Roy, *Special Functions*, Cambridge University Press, Cambridge, 1999.
6. J. Gilewicz, *Approximants de Padé, volume 667 of Lecture Notes in Mathematics*, Springer, 1978.
7. W. B. Jones and W. J. Thron, *Continued Fractions-Analytic Theory and Applications*, volume 11 of Encyclopedia of Mathematics and its Applications, Addison-Wesley, 1980.
8. J. Karamata, *Sur quelques problème posés par Ramanujan*, J. Indian Math. Soc. (N. S.) **24** (1960), 343–365.
9. W. Koepf, *Computer Algebra Algorithms for Orthogonal Polynomials and Special Functions, Leuven 2002*, In Orthogonal Polynomials and Special Functions, volume 1817 of Lecture Notes in Mathematics, Springer, 2003, pp. 1–24.
10. J. H. Lambert, *Memoire sur quelques propri'et's remarquables des quantit'es transcendantes circulaires et logarithmiques*, Memoires de l'Acac. de Berlin, Ann'ee 1761 (1768), 265–322.
11. L. Lorentzen and H. Waadeland, *Continued Fractions with Applications*, volume 3 of Studies in Computational Mathematics, North-Holland, 1992.
12. D. S. Mitrinović, *Analytic Inequalities*, Springer, Berlin, 1970.
13. N. J. A. Sloane, *The On-Line Encyclopedia of Integer Sequences*, Notices of the AMS **50** (2003), 912–915; (The encyclopedia itself is available at http://www.research.att.com/njas/ sequences/).
14. F. G. Tricomi, *Vorlesungen über Ortogonalreihen*, Springer, Berlin Heidelberg, 1995.

In: Inequality Theory and Applications, Volume 4
Editors: Yeol Je Cho et al., pp.153–165

ISBN 1-59454-874-9

ERROR INEQUALITIES FOR AN OPTIMAL 2-POINT QUADRATURE FORMULA OF OPEN TYPE

NENAD UJEVIĆ[1]

ABSTRACT. An optimal 2-point quadrature formula of open type is derived. It is shown that the optimal quadrature formula has a better error bound than the well-known 2-point Gauss quadrature formula. Various error inequalities for this formula are established. Applications in numerical integration are given.

1. Introduction

In recent years, a number of authors have considered an error analysis for quadrature rules of Newton-Cotes type. In particular, the mid-point, trapezoid and Simpson rules have been investigated more recently ([2], [3], [4], [5], [6], [11]) with the view of obtaining bounds on the quadrature rule in terms of a variety of norms involving, at most, the first derivative. Gauss-like quadrature rules are considered in [12] and [14] from an inequalities point of view. These results enlarge the applicability of the mentioned quadrature rules.

In this paper, we derive an optimal 2-point quadrature formula of open type. It is optimal in the sense that it has a minimal error bound. In Section 2 we derive the optimal quadrature formula. We show that this formula has a better estimation of error than the well-known 2-point Gaussian quadrature rule (which is also 2-point quadrature formula of open type). In Section 3, we establish some error bounds for the optimal formula. Similar estimations can be found in [11], [12] and [13], where some different quadrature formulas are considered. In Section 4, we consider an estimation in L_2-norm. These estimations ensure that we can apply the optimal quadrature formula to different classes of functions. In Section 5, we give applications of the above mentioned results in numerical integration.

2. An optimal Quadrature Formula

First, we consider the problem of finding the optimal quadrature formula on the interval $[-1,1]$. Let $\alpha \in R$ and $x, y \in [-1,1]$, $x \leq y$. We define the mapping

$$K(\alpha, x, y, t) = \begin{cases} t+1, & t \in [-1, x), \\ t-\alpha, & t \in [x, y], \\ t-1, & t \in (y, 1]. \end{cases} \tag{1}$$

Received November 17, 2003.
2000*Mathematics Subject Classification*. 26D10, 65D32, 65D30.
Key words and phrases. Optimal quadrature formula, error inequalities, numerical integration.
[1] Department of Mathematics, Univresity of Split, Teslina 12/III, 21000 Split, Croatia. Email: ujevic@pmfst.hr

Let $I \subset R$ be an open interval such that $[-1,1] \subset I$ and let $f : I \to R$ be a differentiable function such that f' is bounded and integrable. We denote

$$\|f\|_\infty = \sup_{t\in[-1,1]} |f(t)| . \tag{2}$$

Integrating by parts, we obtain

$$\int_{-1}^{1} K(\alpha,x,y,t)f'(t)dt = (\alpha+1)f(x) + (1-\alpha)f(y) - \int_{-1}^{1} f(t)dt. \tag{3}$$

We also have

$$\left|\int_{-1}^{1} K(\alpha,x,y,t)f'(t)dt\right| \le \|f'\|_\infty \int_{-1}^{1} |K(\alpha,x,y,t)|\, dt \tag{4}$$

and

$$\int_{-1}^{1} |K(\alpha,x,y,t)|\, dt = \int_{-1}^{x} |t+1|\, dt + \int_{x}^{y} |t-\alpha|\, dt + \int_{y}^{1} |t-1|\, dt. \tag{5}$$

We now define

$$\begin{aligned} g(\alpha,x,y) &= \int_{-1}^{x} |t+1|\, dt + \int_{x}^{y} |t-\alpha|\, dt + \int_{y}^{1} |t-1|\, dt \\ &= \frac{1}{2}\left[(x+1)^2 + (y-1)^2\right] + \int_{x}^{y} |t-\alpha|\, dt \end{aligned} \tag{6}$$

and consider the problem

$$\text{minimize}\, g(p),\ p = (\alpha,x,y) \in R\times[-1,1]\times[-1,1]\,,\ x \le y. \tag{7}$$

Hence we should like to find a global minimizer of $g(p)$ for

$$p \in Q = \{(\alpha,x,y) : \alpha\in R,\ x,y\in[-1,1]\,,\ x\le y\}\,.$$

Recall, a global minimizer is a point $p^* = (\alpha^*, x^*, y^*)$ that satisfies

$$g(p^*) \le g(p), \quad p\in Q. \tag{8}$$

We wish to solve the problem (7). For that purpose, we consider the following cases:

(i) $\alpha \le x$;

(ii) $x \le \alpha \le y$;

(iii) $\alpha \ge y$.

The case (i) We have

$$\begin{aligned} g(\alpha,x,y) &= \frac{1}{2}\left[(x+1)^2 + (y-1)^2\right] + \int_{x}^{y} (t-\alpha)\, dt \\ &= y^2 - y\alpha + x\alpha + x - y + 1. \end{aligned} \tag{9}$$

We seek stationary points of g, defined by (9). For that purpose, we must solve the system:

$$\begin{aligned} \frac{\partial g}{\partial \alpha} &= -y + x = 0, \\ \frac{\partial g}{\partial x} &= \alpha + 1 = 0, \\ \frac{\partial g}{\partial y} &= 2y - \alpha - 1 = 0. \end{aligned}$$

From the above system, we find the stationary point $p_1^* = (-1,0,0)$ and

$$g(p_1^*) = 1. \tag{10}$$

We must also consider the function g on the boundary ∂Q of Q, where $Q = \{(\alpha, x, y) : \alpha \leq x\}$. For $\alpha = x$, we get

$$g(x,x,y) = y^2 + x^2 - xy + x - y + 1$$

and

$$\frac{\partial g}{\partial x} = 2x - y + 1,$$
$$\frac{\partial g}{\partial y} = 2y - x - 1.$$

From the equations $\frac{\partial g}{\partial x} = 0$ and $\frac{\partial g}{\partial y} = 0$, we find the solution $p_2^* = (-\frac{1}{3}, -\frac{1}{3}, \frac{1}{3})$ and

$$g(p_2^*) = \frac{2}{3}. \tag{11}$$

Hence, in this case, we found two candidates for a global minimizer.

The case (iii) We have

$$\begin{aligned} g(\alpha,x,y) &= \frac{1}{2}\left[(x+1)^2 + (y-1)^2\right] - \int_x^y (t-\alpha)\,dt \\ &= x^2 - x\alpha + y\alpha + x - y + 1. \end{aligned} \tag{12}$$

We now seek stationary points of the function g defined by (12). We must solve the system

$$\frac{\partial g}{\partial \alpha} = y - x = 0,$$
$$\frac{\partial g}{\partial x} = 2x - \alpha + 1 = 0,$$
$$\frac{\partial g}{\partial y} = \alpha - 1 = 0.$$

The solution is $p_3^* = (1,0,0)$ and

$$g(p_3^*) = 1. \tag{13}$$

We must also consider the function g on the boundary ∂Q of Q, where $Q = \{(\alpha, x, y) : \alpha \geq y\}$. If we substitute $y = \alpha$ in (12), then we get

$$g(y,x,y) = x^2 + y^2 - xy + x - y + 1.$$

Thus

$$\frac{\partial g}{\partial x} = 2x - y + 1,$$
$$\frac{\partial g}{\partial y} = 2y - x - 1.$$

From the equations $\frac{\partial g}{\partial x} = 0$ and $\frac{\partial g}{\partial y} = 0$, we find the solution $p_4^* = (\frac{1}{3}, -\frac{1}{3}, \frac{1}{3})$ and

$$g(p_4^*) = \frac{2}{3}. \tag{14}$$

Hence, in this case, we also found two candidates for a global minimizer.

The case (ii) We have

$$\begin{aligned} g(\alpha, x, y) &= \frac{1}{2}\left[(x+1)^2 + (y-1)^2\right] - \int_x^\alpha (t-\alpha)\,dt + \int_\alpha^y (t-\alpha)\,dt \\ &= x^2 + y^2 + \alpha^2 - x\alpha - y\alpha + x - y + 1. \end{aligned} \tag{15}$$

We seek stationary points of the function g defined by (15). We get the system

$$\begin{aligned} \frac{\partial g}{\partial \alpha} &= 2\alpha - y - x = 0, \\ \frac{\partial g}{\partial x} &= 2x - \alpha + 1 = 0, \\ \frac{\partial g}{\partial y} &= 2y - \alpha - 1 = 0. \end{aligned} \tag{16}$$

The solution of the above system is $p_5^* = (0, -\frac{1}{2}, \frac{1}{2})$. We have

$$g(p_5^*) = \frac{1}{2}. \tag{17}$$

From (16), we get the Hessian

$$H = \begin{bmatrix} 2 & -1 & -1 \\ -1 & 2 & 0 \\ -1 & 0 & 2 \end{bmatrix}.$$

It is obvious that H is a positive definite matrix. Thus p_5^* is, at least, a local minimizer.

We also consider the function g on the boundary ∂Q of Q, where $Q = \{(\alpha, x, y) : x \le \alpha \le y\}$. For $\alpha = x$, from (15) we get

$$g(x, x, y) = x^2 + y^2 - xy + x - y + 1$$

such that

$$\begin{aligned} \frac{\partial g}{\partial x} &= 2x - y + 1, \\ \frac{\partial g}{\partial y} &= 2y - x - 1. \end{aligned}$$

From the equations $\frac{\partial g}{\partial x} = 0$ and $\frac{\partial g}{\partial y} = 0$, we find $p_6^* = (-\frac{1}{3}, -\frac{1}{3}, \frac{1}{3})$ and

$$g(p_6^*) = \frac{2}{3}. \tag{18}$$

For $\alpha = y$, from (15), we get

$$g(y, x, y) = x^2 + y^2 - xy + x - y + 1$$

such that

$$\frac{\partial g}{\partial x} = 2x - y + 1,$$
$$\frac{\partial g}{\partial y} = 2y - x - 1.$$

From the equations $\frac{\partial g}{\partial x} = 0$ and $\frac{\partial g}{\partial y} = 0$, we find $p_7^* = (\frac{1}{3}, -\frac{1}{3}, \frac{1}{3})$ and

$$g(p_7^*) = \frac{2}{3}. \tag{19}$$

Hence, in this case we found three candidates for a global minimizer.

We considered all possible cases. From (10), (11), (13), (14), (17), (18) and (19), we find that

$$\min_{1 \le i \le 7} g(p_i^*) = g(p_5^*) = \frac{1}{2}.$$

Thus the global minimizer is $p_5^* = (0, -\frac{1}{2}, \frac{1}{2})$. For $\alpha = 0$, $x = -\frac{1}{2}$ and $y = \frac{1}{2}$, from (3), we get

$$\int_{-1}^{1} K\left(0, -\frac{1}{2}, \frac{1}{2}, t\right) f'(t)dt = f\left(-\frac{1}{2}\right) + f\left(\frac{1}{2}\right) - \int_{-1}^{1} f(t)dt. \tag{20}$$

The above quadrature formula is optimal in the sense mentioned in Section 1.

We now summarize the above obtained results.

Theorem 1. *Let $I \subset R$ be an open interval such that $[-1, 1] \subset I$ and let $f : I \to R$ be a differentiable function such that f' is bounded and integrable. Then we have*

$$\left| f\left(-\frac{1}{2}\right) + f\left(\frac{1}{2}\right) - \int_{-1}^{1} f(t)dt \right| \le \frac{\|f'\|_\infty}{2}. \tag{21}$$

Remark 2. The Gaussian 2-point quadrature formula

$$\int_{-1}^{1} K\left(0, -\frac{\sqrt{3}}{3}, \frac{\sqrt{3}}{3}, t\right) f'(t)dt = f\left(-\frac{\sqrt{3}}{3}\right) + f\left(\frac{\sqrt{3}}{3}\right) - \int_{-1}^{1} f(t)dt \tag{22}$$

satisfies

$$\left| f\left(-\frac{\sqrt{3}}{3}\right) + f\left(\frac{\sqrt{3}}{3}\right) - \int_{-1}^{1} f(t)dt \right| \le \frac{\|f'\|_\infty}{3}(5 - 2\sqrt{3}). \tag{23}$$

If we now compare (21) with (23), then we see that the estimate (21) is better than the estimate (23), since $\frac{1}{2} < \frac{5-2\sqrt{3}}{3}$. Hence the quadrature formula (20) has a better estimate of error than the Gaussian quadrature formula (22).

If we consider the above problem on the interval $[a, b]$, then we get the following result.

Theorem 3. *Let $I \subset R$ be an open interval such that $[a,b] \subset I$ and let $f : I \to R$ be a differentiable function such that f' is bounded and integrable. Then we have*

$$\left| \frac{b-a}{2} \left[f\left(\frac{3a+b}{4}\right) + f\left(\frac{a+3b}{4}\right) \right] - \int_a^b f(t)dt \right| \leq \frac{\|f'\|_\infty}{8}(b-a)^2, \tag{24}$$

where $\|f'\|_\infty = \sup_{t\in[a,b]} |f'(t)|$.

3. Error Inequalities

First, we consider some basic properties of the spaces $L_p(a,b)$ for $p = 1, 2, \infty$. As we know, $X = (L_2(a,b), (\cdot,\cdot))$ is a Hilbert space with the inner product

$$(f,g) = \int_a^b f(t)g(t)dt. \tag{25}$$

In the space X, the norm $\|\cdot\|_2$ is defined in the usual way,

$$\|f\|_2 = \left(\int_a^b f(t)^2 dt \right)^{1/2}. \tag{26}$$

We also consider the space $Y = (L_2(a,b), \langle\cdot,\cdot\rangle)$, where the inner product $\langle\cdot,\cdot\rangle$ is defined by

$$\langle f,g \rangle = \frac{1}{b-a} \int_a^b f(t)g(t)dt. \tag{27}$$

It is not difficult to see that Y is a Hilbert space, too. In the space Y, the norm $\|\cdot\|$ is defined by

$$\|f\| = \sqrt{\langle f,f \rangle}. \tag{28}$$

We also define the Chebyshev functional

$$T(f,g) = \langle f,g \rangle - \langle f,e \rangle \langle g,e \rangle, \tag{29}$$

where $f, g \in L_2(a,b)$ and $e = 1$. This functional satisfies the pre-Grüss inequality ([9, p. 296]),

$$T(f,g)^2 \leq T(f,f)T(g,g). \tag{30}$$

Specially, we define

$$\sigma(f) = \sigma(f;a,b) = \sqrt{(b-a)T(f,f)}. \tag{31}$$

The space $L_1(a,b)$ is a Banach space with the norm

$$\|f\|_1 = \int_a^b |f(t)| dt \tag{32}$$

and the space $L_\infty(a,b)$ is also a Banach space with the norm

$$\|f\|_\infty = \operatorname*{esssup}_{t\in[a,b]} |f(t)|. \tag{33}$$

If $f \in L_1(a,b)$ and $g \in L_\infty(a,b)$, then we have

$$|(f,g)| \leq \|f\|_1 \|g\|_\infty. \tag{34}$$

More about the above mentioned spaces can be found, for example, in [1].

Finally, we define the functional

$$\begin{aligned} Q(f) &= Q(f;a,b) \\ &= \int_a^b f(t)dt - \frac{b-a}{2} \left[f\left(\frac{3a+b}{4}\right) + f\left(\frac{a+3b}{4}\right) \right]. \end{aligned} \tag{35}$$

We also need the following lemma.

Lemma 4. *Let*

$$f(t)=\begin{cases} f_1(t), & t\in[a,x_1], \\ f_2(t), & t\in(x_1,x_2], \\ f_3(t), & t\in(x_2,b], \end{cases} \tag{36}$$

where $x_1,x_2\in[a,b]$, $x_1<x_2$, $f_1\in C^1[a,x_1]$, $f_2\in C^1[x_1,x_2]$, $f_3\in C^1[x_2,b]$. *If* $f_1(x_1)=f_2(x_1)$ *and* $f_2(x_2)=f_3(x_2)$, *then* f *is an absolutely continuous function.*

A variant of this lemma can be found in [14].

Theorem 5. *Let* $f:[-1,1]\to R$ *be an absolutely continuous function such that* $f'\in L_1(-1,1)$ *and there exist real numbers* γ_1,Γ_1 *such that* $\gamma_1\le f'(t)\le\Gamma_1$, $t\in[-1,1]$. *Then*

$$|Q(f;-1,1)|\le\frac{\Gamma_1-\gamma_1}{4}, \tag{37}$$

$$|Q(f;-1,1)|\le(S-\gamma_1), \tag{38}$$

$$|Q(f;-1,1)|\le(\Gamma_1-S), \tag{39}$$

where $Q(f;-1,1)$ *is defined by* (35) *and* $S=[f(1)-f(-1)]/2$.

Proof. We define the function

$$p_1(t)=\begin{cases} t+1, & t\in\left[-1,-\frac{1}{2}\right], \\ t, & t\in\left(-\frac{1}{2},\frac{1}{2}\right], \\ t-1, & t\in\left(\frac{1}{2},1\right]. \end{cases} \tag{40}$$

It is easy to verify that

$$(p_1,f')=-Q(f;-1,1). \tag{41}$$

On the other hand, we have

$$\left(f'-\frac{\Gamma_1+\gamma_1}{2},p_1\right)=(f',p_1) \tag{42}$$

since $(p_1,e)=0$. From (34), we get

$$\left|\left(f'-\frac{\Gamma_1+\gamma_1}{2},p_1\right)\right|\le\left\|f'-\frac{\Gamma_1+\gamma_1}{2}\right\|_\infty\|p_1\|_1\le\frac{\Gamma_1-\gamma_1}{4}, \tag{43}$$

since

$$\left\|f'-\frac{\Gamma_1+\gamma_1}{2}\right\|_\infty\le\frac{\Gamma_1-\gamma_1}{2}$$

and

$$\|p_1\|_1=\frac{1}{2}.$$

From (41)-(43), we see that (37) holds. We now prove that (38) holds. We have

$$|(f'-\gamma_1,p_1)|\le\|p_1\|_\infty\|f'-\gamma_1\|_1=(S-\gamma_1)$$

since

$$\|p_1\|_\infty=\frac{1}{2}$$

and

$$\|f'-\gamma_1\|_1=\int_{-1}^{1}(f'(t)-\gamma_1)dt=f(1)-f(-1)-2\gamma_1$$
$$=2(S-\gamma_1).$$

In a similar way, we can prove that (39) holds.

Remark 6. Note that we can apply the estimate (37) only if the first derivative f' is bounded. It means that we cannot use (37) to estimate directly the error when approximating the integral of such a well-behaved function as $f(t) = \sqrt{t}$ on $[0, 1]$, (since $f'(t) = 1/(2\sqrt{t})$ is unbounded on $[0, 1]$). On the other hand, we can use the estimation (38), (since $\gamma_1 = 1/2$ on $[0, 1]$ for the given function).

Remark 7. In [12], we can find the following result for the Gaussian quadrature formula, given in Remark 2,

$$\left| f(-\frac{\sqrt{3}}{3}) + f(\frac{\sqrt{3}}{3}) - \int_{-1}^{1} f(t)dt \right| \leq \frac{\Gamma_1 - \gamma_1}{6}(5 - 2\sqrt{3}). \tag{4.4}$$

We see that (37) is better than (44) since $\frac{1}{4} < \frac{5-2\sqrt{3}}{6}$.

Theorem 8. *Let $f : [a, b] \to R$ be an absolutely continuous function such that $f' \in L_1(a, b)$ and there exist real numbers γ_1, Γ_1 such that $\gamma_1 \leq f'(t) \leq \Gamma_1$, $t \in [a, b]$. Then*

$$|Q(f; a, b)| \leq \frac{\Gamma_1 - \gamma_1}{16}(b - a)^2, \tag{45}$$

$$|Q(f; a, b)| \leq \frac{1}{4}(S - \gamma_1)(b - a)^2, \tag{46}$$

$$|Q(f; a, b)| \leq \frac{1}{4}(\Gamma_1 - S)(b - a)^2, \tag{47}$$

where $Q(f; a, b)$ is defined by (35) and $S = (f(b) - f(a))/(b - a)$.

Theorem 9. *Let $f : [-1, 1] \to R$ be an absolutely continuous function such that $f' \in L_2(-1, 1)$. Then*

$$|Q(f; -1, 1)| \leq \sqrt{\frac{1}{6}}\sigma(f'; -1, 1), \tag{48}$$

where $\sigma(f; -1, 1)$ is defined by (31). The inequality (48) is sharp in the sense that the constant $\sqrt{\frac{1}{6}}$ cannot be replaced by a smaller one.

Proof. Let p_1 be defined by (40). We have

$$\langle p_1, f' \rangle = -\frac{1}{2}Q(f; 0, 1)$$

since (41) holds and $\langle f, g \rangle = \frac{1}{2}(f, g)$ if $[a, b] = [-1, 1]$. On the other hand, we have

$$\langle p_1, f' \rangle = T(f', p_1)$$

since $\langle p_1, e \rangle = 0$. From (30), it follows that

$$|T(f', p_1)| \leq \sqrt{T(p_1, p_1)}\sqrt{T(f', f')} = \frac{1}{2} \|p_1\|_2 \sigma(f'; -1, 1)$$
$$= \frac{1}{2}\sqrt{\frac{1}{6}}\sigma(f'; -1, 1)$$

since

$$\|p_1\|_2 = \sqrt{\frac{1}{6}}.$$

Hence the inequality (48) is proved. We have to prove that this inequality is sharp. For that purpose, we define the function

$$f(t)=\begin{cases}\frac{1}{2}(t+1)^2, & t\in\left[-1,-\frac{1}{2}\right],\\ \frac{1}{2}t^2, & t\in\left(-\frac{1}{2},\frac{1}{2}\right],\\ \frac{1}{2}(t-1)^2, & t\in\left(\frac{1}{2},1\right],\end{cases} \tag{49}$$

such that $f'(t)=p_1(t)$. From Lemma 4, we see that the function f defined by (49) is an absolutely continuous function. For this function, the left-hand side of (48) becomes

$$L.H.S.(48)=\frac{1}{6}.$$

The right-hand side of (48) becomes

$$R.H.S.(48)=\frac{1}{6}.$$

We see that $L.H.S.(48)=R.H.S.(48)$. Thus (48) is sharp.

Theorem 10. *Let $f:[a,b]\to R$ be an absolutely continuous function such that $f'\in L_2(a,b)$. Then*

$$|Q(f;a,b)|\le\frac{1}{4\sqrt{3}}\sigma(f':a,b)(b-a)^{3/2}, \tag{50}$$

where $\sigma(f;a,b)$ is defined by (31). The inequality (50) is sharp in the sense that the constant $\frac{1}{4\sqrt{3}}$ cannot be replaced by a smaller one.

Remark 11. The estimate (45) is better than the estimate (50). However, note that the estimate (45) can be applied only if f' is bounded. On the other hand, the estimate (50) can be applied for an absolutely continuous function if $f'\in L_2(a,b)$.

There are many examples where we cannot apply the estimate (45), but we can apply (50).

Example 12. Let us consider the integral $\int_0^1\sqrt[3]{\sin t^2}dt$. We have

$$f(t)=\sqrt[3]{\sin t^2},\quad f'(t)=\frac{2t\cos t^2}{3\sqrt[3]{\sin^2 t^2}}$$

such that $f'(t)\to\infty,\ t\to 0$ and we cannot apply the estimate (45). On the other hand, we have

$$\int_0^1\left[f'(t)\right]^2dt\le\frac{4}{9}\max_{t\in[0,1]}\frac{t^2\cos t^2}{\sin t^2}\int_0^1\frac{dt}{\sqrt[3]{\sin t^2}}\le\frac{16}{9},$$

i.e., $\|f'\|_2\le\frac{4}{3}$ and we can apply the estimate (50).

4. An Estimation in L_2-norm

Let $K(\alpha,x,y,t)$ be defined by (1) and let f satisfies the assumptions of Theorem 1 with $f'\in L_2(-1,1)$. Then we have

$$\left|\int_{-1}^1K(\alpha,x,y,t)f'(t)dt\right|\le\|f'\|_2\,\|K(\alpha,x,y,\cdot)\|_2\,. \tag{51}$$

We define

$$g(\alpha,x,y)=\int_{-1}^{1}[K(\alpha,x,y,t)]^2\,dt$$

(52)
$$=\int_{-1}^{x}(t+1)^2dt+\int_{x}^{y}(t-\alpha)^2dt+\int_{y}^{1}(t-1)^2dt$$

$$=x^2+y^2-\alpha y^2+\alpha x^2+y\alpha^2-x\alpha^2+x-y+\frac{2}{3}$$

and consider the problem

(53)
$$\text{minimize } g(p),\quad p=(\alpha,x,y)\in R\times[-1,1]^2,\quad x\le y.$$

We seek stationary points of the function g defined by (52). For that purpose, we must solve the system

(54)
$$\begin{aligned}\frac{\partial g}{\partial\alpha}&=x^2-y^2+2y\alpha-2x\alpha=0,\\ \frac{\partial g}{\partial x}&=2x+2x\alpha-\alpha^2+1=0,\\ \frac{\partial g}{\partial y}&=2y-2y\alpha+\alpha^2-1=0.\end{aligned}$$

We can rewrite the above system in the form

(55)
$$\begin{aligned}(y-x)(2\alpha-x-y)&=0,\\ (\alpha+1)(1-\alpha+2x)&=0,\\ (\alpha-1)(1+\alpha-2y)&=0,\end{aligned}$$

which gives the following systems

$$\begin{array}{ccccc} y=x & y=x & x+y=2\alpha & \alpha=1 & 2y=\alpha+1\\ \alpha=-1 & \alpha=1+2x & \alpha=1+2x & \alpha=2x+1 & \alpha=-1\\ \alpha=2y-1 & \alpha=1 & \alpha=2y-1 & 2\alpha=x+y & 2\alpha=x+y.\end{array}$$

We don't consider all systems which can be obtained from (55) since some of them have no solutions.

The stationary points are: $p_1^*=(-1,0,0)$, $p_2^*=(1,0,0)$, $p_3^*=(0,-\frac{1}{2},\frac{1}{2})$, $p_4^*=(1,0,1)$ and $p_5^*=(-1,-1,0)$.

We easily find that the global minimizer is $p_3^*=(0,-\frac{1}{2},\frac{1}{2})$. Let us verify the last fact. From (54), we get the Hessian

$$H(p)=\begin{bmatrix}2y-2x & 2x-2\alpha & 2\alpha-2y\\ 2x-2\alpha & 2\alpha+2 & 0\\ 2\alpha-2y & 0 & 2-2\alpha\end{bmatrix}$$

and

$$H(p_3^*)=\begin{bmatrix}2 & -1 & -1\\ -1 & 2 & 0\\ -1 & 0 & 2\end{bmatrix}.$$

It is obvious that $H(p_3^*)$ is a positive definite matrix. Thus p_3^* is the global minimizer. If we substitute $\alpha=0$, $x=-\frac{1}{2}$ and $y=\frac{1}{2}$ in (47) then we get (20). Hence, we got the following result.

Theorem 13. *Let $I \subset R$ be an open interval such that $[-1,1] \subset I$ and let $f : I \to R$ be a differentiable function such that $f' \in L_2(-1,1)$. Then we have*

$$\left| f\Big(-\frac{1}{2}\Big) + f\Big(\frac{1}{2}\Big) - \int_{-1}^{1} f(t)dt \right| \le \frac{\|f'\|_2}{\sqrt{6}}. \tag{56}$$

We now compare the above obtained result with an estimation for the Gauss 2-point rule. We have

$$\begin{aligned}
&\int_{-1}^{1} \left[K\Big(0, -\frac{\sqrt{3}}{3}, \frac{\sqrt{3}}{3}, t\Big)\right]^2 dt \\
&= \int_{-1}^{-\frac{\sqrt{3}}{3}} (t+1)^2 dt + \int_{-\frac{\sqrt{3}}{3}}^{\frac{\sqrt{3}}{3}} t^2 dt + \int_{\frac{\sqrt{3}}{3}}^{1} (t-1)^2 dt = -\frac{2}{3}\sqrt{3} + \frac{4}{3}.
\end{aligned}$$

Thus

$$\left| f\Big(-\frac{\sqrt{3}}{3}\Big) + f\Big(\frac{\sqrt{3}}{3}\Big) - \int_{-1}^{1} f(t)\, dt \right| \le \sqrt{-\frac{2}{3}\sqrt{3} + \frac{4}{3}}\, \|f'\|_2 \, .$$

Since $\frac{1}{\sqrt{6}} < \sqrt{-\frac{2}{3}\sqrt{3} + \frac{4}{3}}$ we conclude that the optimal quadrature rule (20) has a better estimation of error than the Gaussian rule.

Theorem 14. *Let $I \subset R$ be an open interval such that $[a,b] \subset I$ and let $f : I \to R$ be a differentiable function such that $f' \in L_2(a,b)$. Then we have*

$$\left| \frac{b-a}{2} \left[f\Big(\frac{3a+b}{4}\Big) + f\Big(\frac{a+3b}{4}\Big)\right] - \int_a^b f(t)dt \right| \le \frac{\|f'\|_2}{4\sqrt{3}} (b-a)^{3/2}. \tag{57}$$

Remark 15. If we now compare (24) with (57) then we must note that the estimate (24) can be applied only if f' is bounded, while the estimate (57) can be applied for a function whose derivative is unbounded (if $f' \in L_2(a,b)$).

5. Applications in Numerical Integration

Let $\pi = \{x_0 = a < x_1 < \cdots < x_n = b\}$ be a given subdivision of the interval $[a,b]$ such that $h_i = x_{i+1} - x_i = h = (b-a)/n$. From (35), we get

$$\begin{aligned}
&Q(f; x_i, x_{i+1}) \\
&= \int_{x_i}^{x_{i+1}} f(t)dt - \frac{h}{2}\left[f\Big(\frac{3x_i + x_{i+1}}{4}\Big) + f\Big(\frac{x_i + 3x_{i+1}}{4}\Big)\right].
\end{aligned}$$

If we now sum the above relation over i from 0 to $n-1$, then we get

$$\begin{aligned}
&\sum_{i=0}^{n-1} Q(f; x_i, x_{i+1}) \\
&= \int_a^b f(t)dt - \frac{h}{2} \sum_{i=0}^{n-1} \left[f\Big(\frac{3x_i + x_{i+1}}{4}\Big) + f\Big(\frac{x_i + 3x_{i+1}}{4}\Big)\right].
\end{aligned}$$

We introduce the notation

$$S(f; a, b) = \sum_{i=0}^{n-1} Q(f; x_i, x_{i+1}). \tag{58}$$

We also define

$$\sigma_n(f) = \sum_{i=0}^{n-1} \sqrt{\frac{b-a}{n} \|f'\|_2^2 - [f(x_{i+1}) - f(x_i)]^2} \tag{59}$$

and

$$\omega_n(f) = \left[(b-a)\|f'\|_2^2 - \frac{1}{n}(f(b) - f(a))^2\right]^{1/2}. \tag{60}$$

Theorem 16. *Under the assumptions of Theorem 3, we have*

$$\left|\int_a^b f(t)dt - \frac{h}{2}\sum_{i=0}^{n-1}\left[f\left(\frac{3x_i + x_{i+1}}{4}\right) + f\left(\frac{x_i + 3x_{i+1}}{4}\right)\right]\right|$$
$$\leq \frac{\|f'\|_\infty}{8n}(b-a)^2,$$

where $\{x_0 = a < x_1 < \cdots < x_n = b\}$ *is a uniform subdivision of* $[a, b]$, *i.e.* $x_i = x_0 + ih$, $h = (b-a)/n$, $i = 0, 1, \cdots, n$.

Proof. Apply Theorem 3 to the intervals $[x_i, x_{i+1}]$ and sum.

Theorem 17. *Under the assumptions of Theorem 14, we have*

$$\left|\int_a^b f(t)dt - \frac{h}{2}\sum_{i=0}^{n-1}\left[f\left(\frac{3x_i + x_{i+1}}{4}\right) + f\left(\frac{x_i + 3x_{i+1}}{4}\right)\right]\right|$$
$$\leq \frac{\|f'\|_2}{4\sqrt{3n}}(b-a)^{3/2}.$$

Proof. Apply Theorem 14 to the intervals $[x_i, x_{i+1}]$ and sum.

Theorem 18. *Under the assumptions of Theorem 8 we have*

$$|S(f; a, b| \leq \frac{\Gamma_1 - \gamma_1}{16n}(b-a)^2,$$
$$|S(f; a, b| \leq \frac{S - \gamma_1}{4n}(b-a)^2,$$
$$|S(f; a, b| \leq \frac{\Gamma_1 - S}{4n}(b-a)^2,$$

where $S(f; a, b)$ *is defined by* (58) *and* $\{a = x_0 < x_1 < \cdots < x_n = b\}$ *is a uniform subdivision of* $[a, b]$, *i.e.* $x_i = a + ih$, $h = (b-a)/n$, $i = 0, 1, \cdots, n$.

Proof. Apply Theorem 8 to the intervals $[x_i, x_{i+1}]$ and sum. Note that

$$\sum_{i=0}^{n-1} [f(x_{i+1}) - f(x_i)] = f(b) - f(a).$$

Theorem 19. *Under the assumptions of Theorem 10 we have*

$$|S(f;a,b|\leq\frac{1}{4\sqrt{3}}\frac{b-a}{n}\sigma_n(f)\leq\frac{1}{4\sqrt{3}}\frac{b-a}{\sqrt{n}}\omega_n(f), \tag{61}$$

where $S(f;a,b)$, $\sigma_n(f)$ *and* $\omega_n(f)$ *are defined by* (58), (59) *and* (60), *respectively and* $\{a=x_0<x_1<\cdots<x_n=b\}$ *is a uniform subdivision of* $[a,b]$, *i.e.* $x_i=a+ih$, $h=(b-a)/n$, $i=0,1,\cdots,n$.

Proof. We apply Theorem 10 to the interval $[x_i,x_{i+1}]$ and sum. Then we have

$$\begin{aligned}&|S(f;a,b)|\\&\leq\frac{1}{4\sqrt{3}}h^{3/2}\sum_{i=0}^{n-1}\left[\|f'\|_2^2-\frac{1}{h}\left(f(x_{i+1})-f(x_i)\right)^2\right]^{1/2}.\end{aligned}$$

From the above relation and the fact $h=(b-a)/n$, we see that the first inequality in (61) holds.

Using the Cauchy inequality, we get

$$\begin{aligned}&\sum_{i=0}^{n-1}\left[\|f'\|_2^2-\frac{1}{h}\left(f(x_{i+1})-f(x_i)\right)^2\right]^{1/2}\\&\leq n\left[\|f'\|_2^2-\frac{1}{b-a}\sum_{i=0}^{n-1}\left(f(x_{i+1})-f(x_i)\right)^2\right]^{1/2}\\&\leq n\left[\|f'\|_2^2-\frac{1}{b-a}\frac{1}{n}\left(f(b)-f(a)\right)^2\right]^{1/2}.\end{aligned}$$

Thus the second inequality in (61) holds, too.

References

1. K. Atkinson and W. Han, *Theoretical numerical analysis - A functional analysis framework*, Springer-Verlag, New York/Berlin/Heidelberg, 2001.
2. P. Cerone, *Three points rules in numerical integration*, Nonlinear Anal.-Theory Methods Appl. **47** (2001), 2341–2352.
3. D. Cruz-Uribe and C. J. Neugebauer, *Sharp error bounds for the trapezoidal rule and Simpson's rule*, J. Inequal. Pure Appl. Math. **3** Article 49 (2002), 1–22.
4. S. S. Dragomir, R. P. Agarwal and P. Cerone, *On Simpson's inequality and applications*, J. Inequal. Appl. **5** (2000), 533–579.
5. S. S. Dragomir, P. Cerone and J. Roumeliotis, *A new generalization of Ostrowski's integral inequality for mappings whose derivatives are bounded and applications in numerical integration and for special means*, Appl. Math. Lett. **13** (2000), 19–25.
6. S. S. Dragomir, J. Pečarić and S. Wang, *The unified treatment of trapezoid, Simpson and Ostrowski type inequalities for monotonic mappings and applications*, Math. Comput. Modelling **31** (2000), 61–70.
7. A. Ghizzetti and A. Ossicini, *Quadrature formulae*, Birkhä user Verlag, Basel/Stuttgart, 1970.
8. A. R. Krommer and C.W. Ueberhuber, *Computational integration*, SIAM, Philadelphia, 1998.
9. D. S. Mitrinović, J. Pečarić and A. M. Fink, *Classical and new inequalities in analysis*, Kluwer Acad. Publ., Dordrecht/Boston/London, 1993.
10. S. G. Nash and A. Sofer, *Linear and Nonlinear Programing*, McGraw-Hill Book Co., New York/Singapore, 1996.
11. C. E. M. Pearce, J. Pečarić and N. Ujević and S. Varošanec, *Generalizations of some inequalities of Ostrowski-Grüss type*, Math. Inequal. Appl. **3** (2000), 25–34.
12. N. Ujević, *Inequalities of Ostrowski-Grüss type and applications*, Appl. Math. **29** (2002), 465–479.
13. N. Ujević, *New bounds for the first inequality of Ostrowski-Grüss type*, Comput. Math. Appl. **46** (2003), 421–427.
14. N. Ujević, *Two sharp Ostrowski-like inequalities and applications* (to appearin Meth. Appl. Analysis).

In: Inequality Theory and Applications, Volume 4 ISBN 1-59454-874-9
Editors: Yeol Je Cho et al., pp.167–176

A STRONG INVARIANCE PRINCIPLE FOR QUASI-ASSOCIATED SEQUENCES

WEN SHENG WANG[1]

ABSTRACT. By combining the Csörgő-Révész quantile transform methods and the Skorohod-Strassen martingale embedding theorem, we prove a strong approximation theorem for quasi-associated random variables with mean zero and finite $(2+\delta)$th moment under polynomial decay rate. As a consequence, the decay rate for strong approximation theorem of associated sequences of Yu (1996) is weakened.

1. Introduction

In this paper we present a strong approximation theorem for sums of certain dependent random variables, namely, quasi-associated random variables. quasi-association is a form of positive dependence which is closely related to association and weak association, and it will be informative to see how these various modes of association are related.

Random variables, $X_1, \cdots, X_n$ are *quasi-associated* if, for any $1 \le i \le n-1$ and any two coordinatewise nondecreasing functions $f : \mathbb{R}^i \to \mathbb{R}$ and $g : \mathbb{R}^{n-i} \to \mathbb{R}$,

$$\operatorname{Cov}(f(X_1, \cdots, X_i), g(X_{i+1}, \cdots, X_n)) \ge 0 \tag{1.1}$$

whenever the covariance is defined; are *associated* if, for any two coordinatewise nondecreasing functions $f, g : \mathbb{R}^n \to \mathbb{R}$,

$$\operatorname{Cov}(f(X_1, \cdots, X_n), g(X_1, \cdots, X_n)) \ge 0 \tag{1.2}$$

whenever the covariance is defined; are *weak associated* if, for any $1 \le i \le n-1$ and any two coordinatewise nondecreasing functions $f : \mathbb{R}^i \to \mathbb{R}$ and $g : \mathbb{R}^{n-i} \to \mathbb{R}$ and for any permutation π of positive integers,

$$\operatorname{Cov}(f(X_{\pi(1)}, \cdots, X_{\pi(i)}), g(X_{\pi(i+1)}, \cdots, X_{\pi(n)})) \ge 0 \tag{1.3}$$

whenever the covariance is defined. A sequence $\{X_n, n \ge 1\}$ is quasi-associated (resp., associated, weak associated) if every finite subcollection is quasi-associated (resp., associated, weak associated). It is easy to see that a sequence of association must be weak associated, and that a sequence of weak association must be quasi-associated. So quasi-association is the weakest dependence form of association.

Association was introduced by Esary, Proschan and Walkup [1] and has since found applications in mathematical physics (cf., Newman [2]) and percolation theory (cf., Cox and Grimmett [3]). Weak association was introduced by Burton, Dabrowski and Dehling [4] (in fact, they formulate this concept for collections of $\mathbb{R}^d$-valued random vectors); weak

Received June 2, 2003.
2000*Mathematics Subject Classification.* 60B10, 60F15, 60F17, 62G30.
Key words and phrases. Quasi-association, Partial sum, Strong approximation, Quantile transform.
Research supported by NSFC (10131040).
[1] Department of Mathematics, Hangzhou Teacher's College, Hangzhou 310012, P. R. China. E-mail: wswang@mail.hz.zj.cn

association has applications in reliability theory (cf., Dabrowski and Dehling [5]). Quasi-association was introduced by Khoshnevisan and Lewis [6] in the course of investigating the sample paths of stable processes in random scenery.

For associated sequences, Yu [7], by combining the Berkes-Philipp approximation theorem (cf., Berkes and Philipp [8]) with the Csörgő-Révész quantile transform methods (cf., Csörgő and Révész [9, 10]), investigated a strong approximation theorem for associated random variables with finite $(2+\delta)$th moments. However, because of using the Berkes-Philipp approximation theorem, his result require the exponential decay rate (see Theorem 2.5 therein). An interesting question is weather or not this exponential decay rate can be weakened. One of the purposes of this paper is to weaken this decay rate to polynomial decay rate. But we state the results only for quasi-associated sequences, which also holds for associated sequences because the former dependence form is the weakest dependence form of association as mentioned above.

On the other hand, Lewis [11] was the first, and maybe the unique paper which contains some limit theorems for strictly stationary quasi-associated sequences. In which, a law of the iterated logarithm (LIL) (under finite 3th moment) and a central limit theorem for sums of quasi-associated random variables were established. As another purpose of this paper, we investigate strong approximation theorems for quasi-associated sequences. As consequences, the LIL and Chung's LIL of quasi-associated sequences are obtained. The exact results are stated in Section 2.

In closing this section, let us make some general remarks concerning quasi-association. Quasi-association, like association and weak association, is preserved under certain natural actions on the collection of random variables. We will describe two of these actions: *concatenation* and *ordered blocking*.

Remark 1.1. Let $\mathcal{U}=\{U_1,\cdots,U_m\}$ and $\mathcal{V}=\{V_1,\cdots,V_n\}$ be independent and quasi-associated. Then the collection of random variables formed by cancatenating $\mathcal{U}$ and $\mathcal{V}$, namely

$$\{U_1,\cdots,U_m,V_1,\cdots,V_n\},$$

is quasi-association.

Remark 1.2. Let $\mathcal{U}=\{U_1,\cdots,U_n\}$ be quasi-associated and let $A_1,\cdots,A_k$ be disjoint subsets of $\{1,\cdots,n\}$ with the property that for each integer j, each element of A_j dominates every element of A_{j-1} and dominated, in turn, by each element A_{j+1}. For each integer $1\le j\le k$, let V_j be a nondecreasing function of the random variables $\{U_i, i\in A_j\}$. Then the collection $\mathcal{V}=\{V_1,\cdots,V_k\}$ is quasi-associated as well. We will call the action of passing from $\mathcal{U}$ to $\mathcal{V}$ ordered blocking; thus, quasi-association is preserved under ordered blocking.

2. Main Results

Throughout this paper we will use the following notations: $\{X_n, n\ge 1\}$ be a sequence of random variables on a common probability space; $S_0=0$, $S_n=S(n)=\sum_{i=1}^n X_i, n\ge 1$; $\sigma_n^2=ES^2(n)$; $[x]$ denotes the integer part of x, $a\sim b$ denotes $\lim a/b=1$; $a\ll b$ denotes $a=O(b)$; $\log x=\ln\max(x,e)$ for $x>0$, where ln, as usual, denotes the natural logarithm; $\overset{\text{a.s.}}{=}$ means "$=$" with probability one. Define

$$u(n)=\sup_{k\ge 1}\sum_{j:|j-k|\ge n}\operatorname{Cov}(X_j,X_k) \tag{2.1}$$

the covariance coefficient of a sequence $\{X_n, n\ge 1\}$ (note that $u(n)=2\sum_{j=n}^\infty \operatorname{Cov}(X_1,X_{j+1})$ for stationary X_j).

The next theorem is our principal result.

Theorem 2.1. *Let $\{X_n, n \geq 1\}$ be a sequence of quasi-associated random variables with $EX_n = 0$,*

$$\sup_{n\geq 1} E|X_n|^{2+\delta} < \infty \quad \text{for some } \delta > 0, \tag{2.2}$$

$$\inf_{n\geq 1, k\geq 0} E(S_{n+k} - S_k)^2/n > 0. \tag{2.3}$$

Assume that

$$u(n) \ll n^{-r} \quad \text{for some } r > 5. \tag{2.4}$$

Then, without changing its distribution, we can redefine the sequence $\{X_n, n \geq 1\}$ on a richer probability space together with a standard Wiener process $\{W(t),\ t \in [0, \infty)\}$ such that, for any $d > 0$,

$$S(n) - W(\sigma_n^2) \overset{\text{a.s.}}{=} o\big(\sigma_n (\log n)^{-d}\big) \quad \text{as } n \to \infty. \tag{2.5}$$

From Theorem 2.1, we obtain the following LIL and Chung's LIL easily.

Corollary 2.2. *Under the condition of Theorem 2.1, we have*

$$\limsup_{n\to\infty} \frac{S_n}{(2\sigma_n^2 \log\log \sigma_n^2)^{1/2}} \overset{\text{a.s.}}{=} 1, \tag{2.6}$$

$$\liminf_{n\to\infty} \left(\frac{8\log\log n}{\pi^2 \sigma_n^2}\right)^{1/2} \max_{1\leq i\leq n} |S_i| \overset{\text{a.s.}}{=} 1. \tag{2.7}$$

The proof of Theorem 2.1 will be given in Section 3. In Sections 4 and 5, we prove some propositions and some lemmas.

3. Proof of Theorem 2.1

For $d > 0$, $0 < a < 1/(2+2d)$ and $A > 0$, define blocks H_k and I_k of consecutive positive integers, leaving no gaps between the blocks. The order is $H_1, I_1, H_2, I_2, \cdots$. The length of the blocks are defined by

$$\text{Card}\, H_k = [a(kA)^{a-1} \exp((kA)^a)], \quad \text{Card}\, I_k = [a(kA)^{a-1} \exp((kA)^a/2)], \quad k \geq 1.$$

Set

$$n_k = \sum_{j\leq k} \text{Card}(H_j \cup I_j) \sim \exp((kA)^a),$$

$$u_k = \sum_{j\in H_k} X_j, \quad \lambda_k = Eu_k^2,$$

$$v_k = \sum_{j\in I_k} X_j, \quad \tau_k = Ev_k^2, \quad k \geq 1.$$

Before we use the quantile transform method, we need to introduce a sequence of Gaussian random variables which are used to smooth the sequence $\{u_k, k \geq 1\}$. Let $\{w_k, k \geq 1\}$ be a sequence of independent $N(0, \tau_k^2/2)$-distributed random variables which is also independent of $\{u_k, k \geq 1\}$. Put

$$\xi_k = (u_k + w_k)/(\lambda_k^2 + \tau_k^2/2)^{1/2}, \quad k \geq 1. \tag{3.1}$$

Let F_k denote the distribution function of ξ_k. Note that F_k is continuous since the smooth random variable w_k is used. Now, we can use the quantile transform method to construct a new quasi-associated sequence with Gaussian marginals. Define

$$\eta_k = \Phi^{-1}(F_k(\xi_k)), \quad k \geq 1, \tag{3.2}$$

where Φ^{-1} is the inverse of the standard Gaussian distribution function Φ.

Proposition 3.1. *Under the condition of Theorem 2.1, we have that, for k satisfying $n_k < n \le n_{k+1}$ and any $d > 0$,*

$$\left|S_n - \sum_{i=1}^{k}(\lambda_i^2 + \tau_i^2/2)^{1/2}\eta_i\right| \overset{\text{a.s.}}{=} o\big(\sigma_n(\log n)^{-d}\big) \quad \text{as } n \to \infty. \tag{3.3}$$

Proposition 3.2. *Under the condition of Theorem 2.1, then we can redefine $\{X_i, i \ge 1\}$ on a richer probability space on which there exists a standard Wiener process $\{W(t), t \ge 0\}$ such that, for k satisfying $n_k < n \le n_{k+1}$ and any $d > 0$,*

$$\left|\sum_{i=1}^{k}(\lambda_i^2 + \tau_i^2/2)^{1/2}\eta_i - W(\sigma_n^2)\right| \overset{\text{a.s.}}{=} o\big(\sigma_n(\log n)^{-d}\big) \quad \text{as } n \to \infty. \tag{3.4}$$

From Propositions 3.1 and 3.2, Theorem 2.1 holds immediately. The proofs of Propositions 3.1 and 3.2 will be given in Sections 4 and 5 respectively.

In what follows, we will use C to denote unspecified positive and finite constants whose value may be different in each occurrence.

4. Proof of Proposition 3.1

We need some lemmas. The first lemma verifies that $\{\eta_k, k \ge 1\}$ is a quasi-associated sequence and its covariances are controlled by that of the original sequence.

Lemma 4.1. *Under the condition of Theorem 2.1, we have that the sequence $\{\eta_k, k \ge 1\}$ constructed from (3.2) is quasi-associated with a common distribution $N(0,1)$. Moreover, for any $0 < \theta < 1/2$ and all $i \ne j$,*

$$0 \le E\eta_i\eta_j \le C\big((ijA^2)^{-(a-1)/2}e^{-((iA)^a+(jA)^a)/2}Eu_iu_j\big)^{\theta/(1+\theta)}, \tag{4.1}$$

where C is a positive constant not depending on i, j.

Proof. Since $P(\eta_k \le x) = P(F_k(\xi_k) \le \Phi(x)) = \Phi(x)$, we conclude that each η_k is a standard Gaussian random variable. By Remark 1.2, $u_1, u_2, \cdots$, are quasi-associated, which implies that $\{\xi_k, k \ge 1\}$ constructed from (3.2) is quasi-associated by Remark 1.1. Then $\{\eta_k, k \ge 1\}$ is quasi-associated since $\Phi^{-1}(F_k(\cdot))$ is non-decreasing. So it is enough to show that (4.1) holds. Since (2.3) implies $\lambda_k^2 + \tau_k^2/2 \ge C(kA)^{a-1}e^{(kA)^a}$, we have, by (3.1),

$$E\xi_i\xi_j \le C\big((ijA^2)^{a-1}e^{(iA)^a}e^{(jA)^a}\big)^{-1/2}Eu_iu_j. \tag{4.2}$$

By applying (4.2), along the same lines as that of the proof of Theorem 2.1 of Yu [7], we get (4.1) easily. This completes the proof.

Lemma 4.2. *Under the condition of Theorem 2.1, for integers $k \ge 1$,*

$$\begin{aligned} 0 &\le Eu_iu_{i+k} \\ &\le \begin{cases} C(iA)^{-(r-1)(a-1)}\exp(-(r-1)(iA)^a/2) & \text{if } k = 1,, \\ C(iA)^{-(r-1)(a-1)}(kA)^{-r}\exp(-(r-1)(iA)^a), & \text{if } 2 \le k \le i, \\ C(iA)^{a-1}(kA)^{-r(a-1)}\exp((iA)^a - r(kA)^a)), & \text{if } k > i, \end{cases} \end{aligned} \tag{4.3}$$

and

$$0 \le Ev_iv_{i+k} \le C(iA)^{a-1}(kA)^{-r}\exp\big((iA)^a/2\big). \tag{4.4}$$

Proof. The proof is completely similar to that of Lemma 3.4 of Yu [7].

We are now ready to finish our proof. At first, we assert that (it is easy to verify from the definition of quasi-associated and those proofs of the results stated below) Theorem 1 of Birkel [12], the maximal inequality of Newman and Wright [13], and Newman [2]'s or Burton, Dabrowski and Dehling [4]'s inequality for characteristic functions are all also true for quasi-associated sequences. Now, along the same lines as that of the proof of Theorem 2.4 of Yu [7], by applying Lemmas 4.1 and 4.2, we get Proposition 3.1 easily. So details are omitted.

5. Proof of Proposition 3.2

We start with a useful fact. Let $\hat{u}_0 = 0$, $\hat{u}_i = (\theta_i^2 + \tau_i^2/2)^{1/2}\eta_i$, $\hat{\xi}_i = \hat{u}_i - E(\hat{u}_i|\mathcal{F}_{i-1})$, $\mathcal{F}_i = \sigma(\hat{u}_j, j \le i)$, and $h_i = h(i) = (iA)^{-ad}\exp((iA)^a)$, $i \ge 1$, where $d > 0$. Then $\hat{u}_i$ are Gaussian by Lemma 4.1. The following fact will be used frequently in this section:

Fact 5.1. For $1 \le i < j$, $E(\hat{u}_j|\hat{u}_1 = x_1, \cdots \hat{u}_i = x_i)$ is Gaussian and its distribution is $N(B_{21}B_{11}^{-1}(x_1, \cdots, x_i)', E\hat{u}_j^2 - B_{21}B_{11}^{-1}B_{12})$, where $B_{21} = B_{12}' = (E\hat{u}_1\hat{u}_j, \cdots, E\hat{u}_i\hat{u}_j)$ and B_{11} is the covariance matrix of $(\hat{u}_1, \cdots, \hat{u}_i)$.

In Lemmas 5.2-5.6 below, we all assume that the assumptions of Theorem 2.1 are satisfied.

Lemma 5.2. *For any integer $k \ge 1$ and $0 < \theta < 1/2$,*

$$0 \le E\hat{u}_i\hat{u}_{i+k}$$
$$\le \begin{cases} C(iA)^{\frac{a-1}{1+\theta}(1-r\theta+\theta)}\exp\left(\frac{(iA)^a}{1+\theta}(1-\frac{r\theta}{2}+\frac{\theta}{2}\right), & \text{if } k = 1, \\ C(iA)^{-\frac{a-1}{1+\theta}(\frac{1}{2}+\theta)}(kA)^{-\frac{a-1}{1+\theta}(\frac{1}{2}-r\theta)} & \\ \quad \times\exp\left(\frac{(iA)^a}{1+\theta}(\frac{1}{2}+\theta)+\frac{(kA)^a}{1+\theta}(\frac{1}{2}-r\theta)\right), & \text{if } 2 \le k \le i, \\ C(iA)^{-\frac{a-1}{1+\theta}(1-r\theta+\theta)}(kA)^{-r\frac{\theta}{1+\theta}} & \\ \quad \times\exp\left(\frac{(iA)^a}{1+\theta}(1-r\theta+\theta)\right), & \text{if } k > i. \end{cases} \tag{5.1}$$

Proof. Since (2.2) and (2.3) imply $\lambda_k^2 + \tau_k^2/2 \le C(kA)^{a-1}\exp((kA)^a)$, then, by (4.1) and (4.3), we get (5.1) immediately.

Lemma 5.3. *Let $\mathbb{A}_m$ be the covariance matrix of $(\hat{u}_1, \cdots, \hat{u}_m)$ and $\mathbb{B} = (E\hat{u}_1\hat{u}_n, \cdots, E\hat{u}_m\hat{u}_n)$, $1 \le m < n$. Then*

$$|\mathbb{B}\mathbb{A}_m^{-1}\mathbb{B}'| \le C\left(\sum_{i=1}^{m} E\hat{u}_i\hat{u}_n\right)^2. \tag{5.2}$$

Proof. Let $a_{ij} = E\hat{u}_i\hat{u}_j$ for $i, j = 1, \cdots, m$. Since $a_{ij} \ge 0$ and $E\hat{u}_i^2 = \theta_i^2 + \tau_i^2/2 \ge C(iA)^{a-1}\exp((iA)^a)$, it is easy to verify that

$$\sum_{j\ge1;\, j\ne i} |a_{ij}| = \sum_{j\ge1;\, j\ne i} a_{ij} = o(E\hat{u}_i^2), \quad 1 \le i \le m,$$

by (5.1) (with choosing θ such that $1 - \frac{r\theta}{2} + \frac{\theta}{2} < 0$ therein) and taking $A > 0$ sufficiently large. It follows that $\mathbb{A}_m$ is a strictly dominant principal diagonal, that is, a matrix has positive diagonal elements and the sum of the absolute values of all the off-diagonal elements in a given row is less than the diagonal element in that row. Hence, by Theorem 1 of Price [14], we have

$$|\mathbb{A}_m| \sim \prod_{i=1}^{m} E\hat{u}_i^2. \tag{5.3}$$

Let $\mathbb{A}_m^{(ij)}$ be the submatrix of $\mathbb{A}_m$ obtaining by deleting the ith row and j^{th} column, and let $\sigma_{ij} \stackrel{\text{def}}{=} (-1)^{i+j}|\mathbb{A}_m^{(ij)}|/|\mathbb{A}_m|$. Then it is known that σ_{ij} are elements of $\mathbb{A}_m^{-1}$ and, by (5.3),

$$|\sigma_{ij}| = O(1), \tag{5.4}$$

which implies the desired result.

Lemma 5.4. *For any $d > 0$, we have*

$$\sum_{i=1}^{k} E(\hat{u}_i|\mathcal{F}_{i-1}) \stackrel{\text{a.s.}}{=} o(h_k). \tag{5.5}$$

Proof. Note that

$$E\left(\sum_{i=1}^{k} E(\hat{u}_i|\mathcal{F}_{i-1})\right)^2 = \sum_{i=1}^{k} E(E(\hat{u}_i|\mathcal{F}_{i-1}))^2 + 2\sum_{i=1}^{k}\sum_{j=i}^{k} E(\hat{u}_j E(\hat{u}_i|\mathcal{F}_{i-1})). \tag{5.6}$$

Let $\mathbb{A}_n$, $\mathbb{A}_n^{(ij)}$ be as above. Since $\hat{u}_1, \cdots, \hat{u}_i$ are Gaussian, it is known that the conditional variance can be written as

$$\mathrm{Var}(\hat{u}_j|\hat{u}_i, i \neq j, 1 \leq i \leq n) = \frac{|\mathbb{A}_n|}{|\mathbb{A}_n^{(jj)}|}.$$

Thus

$$\mathrm{Var}(\hat{u}_i|\hat{u}_1, \cdots, \hat{u}_{i-1}) = \frac{|\mathbb{A}_i|}{|\mathbb{A}_{i-1}|} = |E\hat{u}_i^2 - \mathbb{B}\mathbb{A}_{i-1}^{-1}\mathbb{B}'|, \tag{5.7}$$

where $\mathbb{B} = (E\hat{u}_1\hat{u}_i, \cdots, E\hat{u}_{i-1}\hat{u}_i)$. Since $E(E(\hat{u}_i|\mathcal{F}_{i-1}))^2 = E\hat{u}_i^2 - \mathrm{Var}(\hat{u}_i|\hat{u}_1, \cdots, \hat{u}_{i-1})$, then, by (5.7), Lemmas 5.2 and 5.3, we get

$$E(E(\hat{u}_i|\mathcal{F}_{i-1}))^2 \leq |\mathbb{B}\mathbb{A}_{i-1}^{-1}\mathbb{B}'| \leq C\left(\sum_{l=1}^{i-1} E\hat{u}_l\hat{u}_i\right)^2 = O(1). \tag{5.8}$$

Moreover, by Fact 5.1, for $j > i$,

$$\begin{aligned} &E(\hat{u}_j E(\hat{u}_i|\mathcal{F}_{i-1})) \\ &= \int_{\mathbb{R}^i} u_j \left(\int_{\mathbb{R}} u_i \frac{p(u_1, \cdots, u_i)}{p(u_1, \cdots, u_{i-1})} du_i\right) p(u_1, \cdot s u_{i-1}, u_j) du_1 \cdots du_{i-1} du_j \\ &= \int_{\mathbb{R}^i} u_j \{B_{21}B_{11}^{-1}(u_1, \cdots, u_{i-1})'\} p(u_1, \cdots u_{i-1}, u_j) du_1 \cdots du_{i-1} du_j \\ &= B_{21}B_{11}^{-1}(E\hat{u}_1\hat{u}_j, \cdots, E\hat{u}_{i-1}\hat{u}_j)', \end{aligned} \tag{5.9}$$

where B_{11} is the covariance matrix of $(\hat{u}_1, \cdots, \hat{u}_{i-1})$, $B_{21} = (E\hat{u}_1\hat{u}_i, \cdots E\hat{u}_{i-1}\hat{u}_i)$ and $p(u_1, \cdots, u_k)$ is joint density function of $(\hat{u}_1, \cdots, \hat{u}_k)$. Therefore, by (5.1) and the argument of (5.4),

$$\begin{aligned}\sum_{i=1}^{k}\sum_{j=i}^{k} E(\hat{u}_j E(\hat{u}_i|\mathcal{F}_{i-1})) &= \sum_{i=1}^{k}\sum_{j=i}^{k} B_{21}B_{11}^{-1}(E\hat{u}_1\hat{u}_j, \cdots, E\hat{u}_{i-1}\hat{u}_j)' \\ &\ll \sum_{i=1}^{k}\sum_{j=i}^{k}\sum_{l=1}^{i-1} E(\hat{u}_l\hat{u}_i)E(\hat{u}_l\hat{u}_j) \ll k^2.\end{aligned} \tag{5.10}$$

By (5.6), (5.8) and (5.10), we obtain

$$E\left(\sum_{i=1}^{k} E(\hat{u}_i|\mathcal{F}_{i-1})\right)^2 \ll k^2,$$

which, via Borel-Cantelli lemma, implies (5.5) immediately.

Lemma 5.5. *For any $d > 0$, we have*

$$\sum_{i=1}^{\infty} h_i^{-2} E|\hat{\xi}_i|^4 < \infty. \tag{5.11}$$

Proof. Note that

$$E|\hat{\xi}_i|^4 \le CE|\hat{u}_i|^4 \le C(\theta_i^2 + \tau_i^2/2)^2 E|\eta_i|^4 \le C((iA)^{a-1}e^{(iA)^a})^2,$$

which implies (5.11) easily since $a < 1/(2+2d)$.

Lemma 5.6. *For any $d > 0$, we have*

$$\sum_{i=1}^{k} E(\hat{\xi}_i^2|\mathcal{F}_{i-1}) - E\hat{\xi}_i^2 \overset{\text{a.s.}}{=} o(h_k). \tag{5.12}$$

Proof. Note that

$$\begin{aligned}&\left|\sum_{i=1}^{k} E(\hat{\xi}_i^2|\mathcal{F}_{i-1}) - E\hat{\xi}_i^2\right| \\ &\le \left|\sum_{i=1}^{k} E(\hat{u}_i^2|\mathcal{F}_{i-1}) - E\hat{u}_i^2\right| + \sum_{i=1}^{k}(E(\hat{u}_i|\mathcal{F}_{i-1}))^2 + E(E(\hat{u}_i|\mathcal{F}_{i-1}))^2.\end{aligned}$$

By (5.8), we have

$$\sum_{i=1}^{\infty} h_i^{-1} E(E(\hat{u}_i|\mathcal{F}_{i-1}))^2 < \infty.$$

By Kronecker lemma, this proves

$$\sum_{i=1}^{k}(E(\hat{u}_i|\mathcal{F}_{i-1}))^2 + E(E(\hat{u}_i|\mathcal{F}_{i-1}))^2 \overset{\text{a.s.}}{=} o(h_k).$$

Hence it is enough to show that

$$\sum_{i=1}^{k} E(\hat{u}_i^2|\mathcal{F}_{i-1}) - E\hat{u}_i^2 \overset{\text{a.s.}}{=} o(h_k). \tag{5.13}$$

By Fact 1 (similarly, to (5.9)),

$$\begin{aligned}
&E(\hat{u}_i^2 E(\hat{u}_i^2|\mathcal{F}_{i-1})) \\
&= \int_{\mathbb{R}^i} u_i^2 \left(\int_{\mathbb{R}} u_i^2 \frac{p(u_1,\cdots,u_i)}{p(u_1,\cdots,u_{i-1})} du_i \right) p(u_1,\cdots u_{i-1},u_i) du_1 \cdots du_{i-1} du_i \\
&= \int_{\mathbb{R}^i} u_i^2 \{E\hat{u}_i^2 - B_{21}B_{11}^{-1}B_{12} + (B_{21}B_{11}^{-1}(u_1,\cdots,u_{i-1})')^2\} \\
&\quad \times p(u_1,\cdots u_{i-1},u_i) du_1 \cdots du_{i-1} du_i \\
&= (E\hat{u}_i^2)^2 - E\hat{u}_i^2(B_{21}B_{11}^{-1}B_{12}) + E\{\hat{u}_i(B_{21}B_{11}^{-1}(\hat{u}_1,\cdots,\hat{u}_{i-1})')\}^2,
\end{aligned}$$

where $p(\cdot)$, B_{11}, B_{21} and B_{12} are defined as in (5.9). Hence

$$\begin{aligned}
&E(E(\hat{u}_i^2|\mathcal{F}_{i-1}) - E\hat{u}_i^2))^2 \\
&= (E\hat{u}_i^2)^2 - E(\hat{u}_i^2 E(\hat{u}_i^2|\mathcal{F}_{i-1})) \\
&= E\hat{u}_i^2(B_{21}B_{11}^{-1}B_{12}) - E\{\hat{u}_i(B_{21}B_{11}^{-1}(\hat{u}_1,\cdots,\hat{u}_{i-1})')\}^2 \\
&\le E\hat{u}_i^2|B_{21}B_{11}^{-1}B_{12}|.
\end{aligned}$$

Thus, by Lemma 5.3,

$$\begin{aligned}
E\left(\sum_{i=1}^{k} E(\hat{u}_i^2|\mathcal{F}_{i-1}) - E\hat{u}_i^2\right)^2 &\le k \max_{1\le i\le k} E\hat{u}_i^2|B_{21}B_{11}^{-1}B_{12}| \le Ck^4 E\hat{u}_k^2 \\
&\le Ck^4k^{a-1}\exp((kA)^a),
\end{aligned}$$

which, via Borel-Cantelli lemma, implies that (5.13) holds. This proves the desired result.

Proof of Proposition 3.2. It follows from the Skorohod-Strassen embedding theorem (cf., Hall and Heyde [15], pp. 269) that on a possibly enlarge version of the underlying probability space there exist a standard Wiener process $\{W(t), t \ge 0\}$ and a non-negative sequence $\tau_n, n \ge 1$ such that

$$\sum_{i=1}^{k} \hat{\xi}_i \overset{\text{a.s.}}{=} W\left(\sum_{i=1}^{k} \tau_i\right), \quad k \ge 1.$$

Furthermore, there exists an increasing sequence $\{\mathcal{A}_n, n \ge 1\}$ of σ-fields such that τ_n is $\mathcal{A}_n$-measurable and that

$$E(\tau_n|\mathcal{A}_{n-1}) \overset{\text{a.s.}}{=} E(\hat{\xi}_n^2|\mathcal{F}_{n-1})$$

and, for $p > 1$,

$$E(\tau_n^p|\mathcal{A}_{n-1}) \overset{\text{a.s.}}{\le} C_p E(|\hat{\xi}_n|^{2p}|\mathcal{F}_{n-1}).$$

Hence

$$\sum_{i=1}^{k}(\tau_i - E\tau_i) = \sum_{i=1}^{k}(\tau_i - E(\tau_i|\mathcal{A}_{i-1})) + \sum_{i=1}^{k}(E(\hat{\xi}_i^2|\mathcal{F}_{i-1}) - E\hat{\xi}_i^2).$$

By Lemmas 5.5 and 5.6 and the well-known almost sure convergence for martingale difference sequence (cf., Chow [16]), we have

$$\sum_{i=1}^{k}(\tau_i - E\tau_i) \overset{\text{a.s.}}{=} o(h_k). \tag{5.14}$$

Noting that

$$\sum_{i=1}^{k}\hat{\xi}_i - W\left(E\left(\sum_{i=1}^{k}\hat{\xi}_i\right)^2\right) \overset{\text{a.s.}}{=} W\left(\sum_{i=1}^{k}\tau_i\right) - W\left(\sum_{i=1}^{k}E\tau_i\right),$$

by (5.14) and Theorem 3.2A of Hanson and Russo [17], we conclude that

$$\sum_{i=1}^{k}\hat{\xi}_i - W(b_k) \overset{\text{a.s.}}{=} o\left(\left(h_k\left(\log\frac{b_k}{h_k} + \log\log h_k\right)\right)^{1/2}\right),$$

where $b_k = \sum_{i=1}^{k} E\hat{\xi}_i^2$. Hence, by Lemma 5.4, it is enough to prove that, for $n_k \le n < n_{k+1}$,

$$W(b_k) - W\left(\sum_{i=1}^{k}(\theta_i^2 + \tau_i^2/2)\right) \overset{\text{a.s.}}{=} o(h_k), \tag{5.15}$$

$$W\left(\sum_{i=1}^{k}(\theta_i^2 + \tau_i^2/2)\right) - W(\sigma_n^2) \overset{\text{a.s.}}{=} o(h_k). \tag{5.16}$$

(5.16) follows by applying Theorem 2.1 of Csörgő and Révész [18] and the inequality

$$0 \le \sigma_n^2 - \sum_{i=1}^{k}(\theta_i^2 + \tau_i^2/2) = o(h_k).$$

To prove (5.15), by Theorem 3.2A of Hanson and Russo [17], it is enough to prove that

$$b_k - \sum_{i=1}^{k}(\theta_i^2 + \tau_i^2/2) \overset{\text{a.s.}}{=} o(h_k). \tag{5.17}$$

Since

$$b_k = \sum_{i=1}^{k} E\hat{u}_i^2 - \sum_{i=1}^{k} E(\hat{u}_i E(\hat{u}_i|\mathcal{F}_{i-1}))$$

and

$$E(\hat{u}_i E(\hat{u}_i|\mathcal{F}_{i-1})) = o(k) \quad \text{for } 1 \le i \le k$$

by the argument of (5.9), then we conclude (5.17) immediately. This completes the proof.

References

1. J. Esary, F. Proschan and D. Walkup, *Association of random variables with applications*, Ann. Math. Statist. **38** (1967), 1466–1474.
2. C. M. Newman, *Normal fluctuations and the FKG inequalities*, Comm. Math. Phys. **74** (1980), 119–128.
3. J. T. Cox and G. Grimmett, *Central limit theorems for associated random variables and the percolation model*, Ann. Probab. **12** (1984), 514–528.

4. R. M. Burton, A. R. Dabrowski and H. Dehling, *An invariance principle for weakly associated random variables*, Stoch. Process. Appl. **23** (1986), 301–306.
5. A. R. Dabrowski and H. Dehling, *A Berry-Esseen theorem and a functional law of the iterated logarithm for weakly associated random variables*, Stoch. Process. Appl. **30** (1988), 277–289.
6. D. Khoshnevisan and T. M. Lewis, *A law of the iterated logarithm for stable processes in random scenery*, Stoch. Process. Appl. **74** (1998), 89–121.
7. H. Yu, *A strong invariance principle for associated sequences*, Ann. Probab. **24** (1996), 2079–2097.
8. I. Berkes and W. Philipp, *Approximation theorems for independent and weakly dependence random variables*, Ann. Probab. **7** (1979), 29-54.
9. M. Csörgő and P. Révész, *A new method to prove Strassen type laws of invariance principle, I*, Z. Wahrsch. Verw. Gebiete **31** (1975), 255–260.
10. ———, *A new method to prove Strassen type laws of invariance principle II*, Z. Wahrsch. Verw. Gebiete **31** (1975), 261–269.
11. T. M. Lewis, *Limit theorems for partial sums of quasi-associated random variables*, Asymptotic Methods in Probability and Statistics (1998), 31–48.
12. T. Birkel, *Moment bounds for associated sequences*, Ann. Probab. **16** (1988), 1184–1193.
13. C. M. Newman and A. L. Wright, *An invariance principle for certain dependent sequences*, Ann. Probab. **9** (1981), 671–675.
14. G. B. Price, *Bounds for determinants with dominant principle diagonal*, Proc. Amer. Math. Soc. **2** (1951), 497–502.
15. P. Hall and C. C. Heyde, *Martingale Limit Theory and its Application*, Academic Press, New York, 1980.
16. Y. S. Chow, *Local convergence of martingale and the law of large numbers*, Ann. Math. Statist. **36** (1965), 552–558.
17. D. L. Hanson, Russo and P. Ralph, *Some results on increments of the Wiener processes with applications to lag sums of iidrv*, Ann. Probab. **11** (1983), 609–623.
18. M. Csörgő and P. Révész, *Strong Approximations in Probability and Statistics*, Academic Press, New York, 1981.

In: Inequality Theory and Applications, Volume 4
Editors: Yeol Je Cho et al., pp.177–183

ISBN 1-59454-874-9

PARETO OPTIMAL TRAJECTORIES IN A MODEL WITH DISCRETE INNOVATIONS

ALEXANDER J. ZASLAVSKI[1]

ABSTRACT. In this paper we consider a model of economic dynamics with discrete innovations and establish the existence of Pareto optimal trajectories.

1. Introduction

We consider the model with discrete innovations introduced in [1] and studied in [2-9]. In this model the state of the economy is determined by a set of operating technologies, a collection of funds corresponding to these technologies, and a set of known, but as yet not implemented technologies. To introduce a new technology, expenditures of the already utilized types of funds are required. As a result of these expenditures, the new technology at the next instant of time will be introduced into action with a certain initial reserve of new funds.

We consider a single-product economy which deals with two production factors: labor L and funds K. The time is assumed to be discrete and the amount of labor is constant and equal to unity. The state of the economy is determined by a set of operating technologies, a collection of funds corresponding to these technologies and a set of known, but as yet not implemented technologies.

A technology is a pair (f, v) where f is a production function of two variables K, L and $v \in [0, 1)$. Possessing at time t funds K and labor resources L, the economy utilizing the technology (f, v) will produce during a unit time interval, a product in the amount of $f(K, L)$. Moreover, at time $t + 1$ the economy will still have in its possession the used old funds in the amount of vK.

To introduce a new technology, expenditures of the already utilized types of funds are required. As a result of these expenditures, the new technology at the next instant of time will be introduced into action with a certain intial reserve of the new funds.

Let $I = \{0, 1, \cdots\}$ and $\{(f^i, v^i) : i \in I\}$ be the set of all technologies which can be utilized in the production process. At time $t \in I$ the state of the economy is given in the form

$$(I_0^t, I_n^t, (K_t^i, C_t^i)(i \in I_0^t)),$$

where I_0^t is a finite set of numbers (indices) of technologies introduced by the time t, I_n^t is the set of numbers of technologies which are available in principle but not introduced,

---

Received February 23, 2004.
1991 *Mathematics Subject Classification.* 91B62.
Key Words and Phrases. Discrete innovation, Pareto optimal trajectory, technology.

[1] Department of Mathematics, The Technion-Israel Institute of Technology, 32000 Haifa, Israel. E-mail: ajzasl@tx.technion.ac.il

and $K_t^i, C_t^i \geq 0$ are the funds and consumption of the i-th type available at time t which correspond to the technology with the number i. We will assume that at time t the following information is available

$$(K^i),\ (s^{ij}) \quad (i \in I_n^t,\ j \in I_n^t \cup I_0^t),$$

where $s^{ij} \geq 0$ is the expenditure of the i-th funds required for introduction of the i-th technology and $K^i > 0$ is the initial amount of the i-th fund which is obtained at the initial time of utilization of the i-th technology. At time $t+1$ the economy may pass over to the state

$$(I_0^{t+1}, I_n^{t+1}, (K_{t+1}^i, C_{t+1}^i)\ (i \in I_0^{t+1}))$$

for which

$$I_0^t \subset I_0^{t+1} \subset I_0^t \cup I_n^t,$$
$$I_0^{t+1} \setminus I_0^t \subset \{i \in I_n^t :\ s^{ij} = 0 \text{ for all } j \in I_n^t\}$$

and the numbers $L_t^i \geq 0$, $i \in I_0^t$ are determined such that

$$\sum_{i \in I_0^t} L_t^i \leq 1,$$
$$K_{t+1}^j \geq v^j K_t^j, \quad j \in I_0^t,$$
$$K_{t+1}^i = K^i,\ C_{t+1}^i = 0, \quad i \in I_0^{t+1} \setminus I_0^t,$$
$$K_{t+1}^j - v^j K_t^j + C_{t+1}^j + \sum_{i \in I_0^{t+1} \setminus I_0^t} s^{ij} \leq f^j(K_t^j, L_t^j), \quad j \in I_0^t.$$

(We assume here that the result of a summation over an empty set equals zero.) Note that in the model under consideration the newly procuced product is used for consumption and expenditures related to an introduction of new technologies. Sometimes the state of the economy at time t will be written in the form

$$(I_0^t, I_n^t, (K_t^i, C_t^i, L_t^i)\ (i \in I_0^t))$$

where (L_t^i) $(i \in I_0^t)$ is the distribution of labor resources at time t. We do it in the case when some description of this distribution is required. When describing a trajectory of the model we would also include into its description the corresponding sequence of distributions of labor resources, most often only in the case when some information about these resources is required. However, in any case a definite sequence of distributions of labor resources is always associated with a trajectory of the model.

Denote by R_+^l the cone of elements of the Euclidean space R^l with nonnegative coordinates. Below all the technologies under consideration (f, v) will assume to be such that $f : R_+^2 \to R_+$ be a continuous, superlinear (superadditive, positively homogeneous) function,

$$f(0,1) = f(1,0) = 0,$$

$$f(x,1) < f(\lambda x, 1) < \lambda f(x,1) \text{ for each } \lambda > 1 \text{ and each } x > 0,$$

and there exists $X \in R_+$ such that $f(1, X) > 1 - v$.

Let (f, v) be a technology. It is easy to see that there exists a unique number $x(f, v) > 0$ such that

$$f(x(f,v), 1) = (1 - v)x(f, v).$$

For $x_0 > 0$ the inequality $f(x_0, 1) > (1 - v)x_0$ holds if and only if $x_0 < x(f, v)$ and the sequence

$$x_t = vx_{t-1} + f(x_{t-1}, 1),\ t = 1, 2, \cdots,$$

converges to $x(f, v)$ as $t \to \infty$. Evidently $x(f, v)$ is a characteristic of the technology (f, v) which evaluates its production capabilities.

The technology (f, v) is associated with a dynamic model of the economy whose trajectory is a sequence (K_t, C_t), $t = 0, 1, \cdots$, where $K_t, C_t \geq 0$ are the funds and consumption available at time t which satisfy

$$K_{t+1} - vK_t \geq 0$$

and

$$K_{t+1} - vK_t + C_{t+1} \leq f(K_t, 1)$$

for all $t = 0, 1, \cdots$.

It is easy to see that, for any model trajectory (K_t, C_t), $t = 0, 1, \cdots$, we have

$$\limsup_{t\to\infty}(vK_t + f(K_t, 1)) \leq x(f, v)$$

and

$$\limsup_{t\to\infty} K_t \leq x(f, v).$$

Moreover, for any initial state of the model (K_0, C_0) with $K_0 > 0$, there exists a trajectory (K_t, C_t), $t = 0, 1, \cdots$ such that $K_t \to x(f, v)$ as $t \to \infty$.

Let $X = (K_t, C_t)(t \in I)$ be a model trajectory. Set

$$w(X) = \limsup_{T\to\infty} T^{-1} \sum_{t=0}^{T-1} C_t.$$

Evidently, $w(X) \leq x(f, v)$. Set

$$w(f, v) = \sup\{w(X) :\ X \text{ is a model trajectory}\}.$$

The number $w(f, v)$ is a characteristic of the technology (f, v) which evaluates its consumption capabilities.

It is easy to verify that the following result is true.

Proposition 1.1. *There exists a number $h(f, v) \in (0, 1)$ such that*

$$\lim_{x\to\infty} (h(f, v)f(1, x)) \in (1 - v, \infty]$$

and

$$(1 - h(f, v))f(x(h(f, v)f, v), 1) = w(f, v) > 0.$$

2. The Main Result

We consider the model with discrete innovations introduced in Section 1. Assume that there is a natural number Q_* such that for each state of the economy

$$(I_0^t, I_n^t, (K_t^i, C_t^i)(i \in I_0^t))$$

at any instant of time t the following inequality holds:

$$\sup\{i :\ i \in I_n^t\} \leq \sup\{i \in:\ i \in I_0^t\} + Q_*. \tag{2.1}$$

We use the following notation. Let

$$Y = (I_0^t, I_n^t, (K_t^i, C_t^i, L_t^i)(i \in I_0^t)) \quad (t \in I)$$

be a model trajectory. For each $t \in I$ and each $i \in I_0^t$ we set

$$I_0^t(Y) = I_0^t,\ I_n^t(Y) = I_n^t,$$

$$K_t^i(Y) = K_t^i,\ C_t^i(Y) = C_t^i,\ L_t^i(Y) = L_t^i \tag{2.2}$$

and

$$C_t(Y) = \sum_{i \in I_0^t} C_i^t,\ K_t(Y) = \sum_{i \in I_0^t} K_i^t.$$

Let

$$X = (I_0^t, I_n^t, (K_t^i, C_t^i)(i \in I_0^t)) \quad (t \in I)$$

be a model trajectory. We say that X is a Pareto optimal trajectory if for each trajectory

$$\tilde{X} = (\tilde{I}_0^t, \tilde{I}_n^t, (\tilde{K}_t^i, \tilde{C}_t^i)(i \in \tilde{I}_0^t)) \quad (t \in I)$$

which satisfies

$$(\tilde{I}_0^0, \tilde{I}_n^0, (\tilde{K}_0^i, \tilde{C}_0^i)(i \in \tilde{I}_0^0)) = (I_0^0, I_n^0, (K_0^i, C_0^i)(i \in I_0^0))$$

and

$$\sum_{i \in \tilde{I}_0^t} \tilde{C}_t^i \geq \sum_{i \in I_0^t} C_t^i,\ t = 0, 1, \cdots,$$

the following equality holds:

$$\sum_{i \in \tilde{I}_0^t} \tilde{C}_t^i = \sum_{i \in I_0^t} C_t^i,\ t = 0, 1, \cdots.$$

The following theorem in the main result of the paper.

Theorem 2.1. *Let*

$$X = (I_0^t, I_n^t, (K_t^i, C_t^i)(i \in I_0^t)) \quad (t \in I)$$

be a model trajectory. Then there exists a Pareto optimal trajectory

$$\tilde{X} = (\tilde{I}_0^t, \tilde{I}_n^t, (\tilde{K}_t^i, \tilde{C}_t^i)(i \in \tilde{I}_0^t)) \quad (t \in I)$$

such that

$$(\tilde{I}_0^0, \tilde{I}_n^0, (\tilde{K}_0^i, \tilde{C}_0^i)(i \in \tilde{I}_0^0)) = (I_0^0, I_n^0, (K_0^i, C_0^i)(i \in I_0^0))$$

and that for each $t \in I$

$$\sum_{i \in \tilde{I}_0^t} \tilde{C}_t^i \geq \sum_{i \in I_0^t} C_t^i.$$

3. Proof of Theorem 2.1

Denote by $\mathfrak{A}$ the set of all trajectories

$$Y = (I_0^t(Y), I_n^t(Y), (K_t^i(Y), C_t^i(Y))(i \in I_0^t(Y))) \quad (t \in I)$$

such that

$$\begin{aligned}&(I_0^0(Y), I_n^0(Y), (K_0^i(Y), C_0^i(Y))(i \in I_0^0(Y)))\\&= (I_0^0, I_n^0, (K_0^i, C_0^i)(i \in I_0^0)).\end{aligned}$$

Let $Y_1, Y_2 \in \mathfrak{A}$. We say that $Y_1 \leq Y_2$ if for each natural number t we have

$$C_t(Y_1) \leq C_t(Y_2). \tag{3.1}$$

We will preface the proof of Theorem 2.1 by the following auxiliary result.

Lemma 3.1. *Assume that $q \in I$ and $Y_0 \in \mathfrak{A}$ satisfy $Y_0 \geq X$ and has the following property:*
(C1) *If $Z \in \mathfrak{A}$ and $Z \geq Y_0$, then*

$$C_t(Y_0) = C_t(Z),\ t = 0, \cdots, q.$$

Then there exists $Y_1 \in \mathfrak{A}$ such that $Y_1 \geq Y_0$ and the following property holds:
(C2) *If $Z \in \mathfrak{A}$ and $Z \geq Y_1$, then*

$$C_t(Y_1) = C_t(Z),\ t = 0, \cdots, q+1.$$

Proof. In view of (2.1), the set

$$\{C_{q+1}(Z) : Z \in \mathfrak{A}\}$$

is bounded. Set

$$M = \sup\{C_{q+1}(Z) : Z \in \mathfrak{A} \text{ and } Z \geq Y_0\}. \tag{3.2}$$

There exists a sequence of model trajectories

$$\{Z_m\}_{m=1}^{\infty} \subset \{Z \in \mathfrak{A} : Z \geq Y_0\}$$

such that

$$C_{q+1}(Z_m) \to M \text{ as } m \to \infty.$$

Using diagonalization process and extracting a subsequence if necessary, we may assume that

$$I_0^t(Z_m) = I_0^t(Z_1),\ m = 1, 2, \cdots,$$
$$I_n^t(Z_m) = I_n^t(Z_1),\ m = 1, 2, \cdots,$$

and, for each integer $t \geq 1$ and each $i \in I_0^t(Z_1)$, there exist

$$\lim_{m\to\infty} K_t^i(Z_m),\quad \lim_{m\to\infty} C_t^i(Z_m),\quad \lim_{m\to\infty} L_t^i(Z_m).$$

Define a model trajectory Y_1 as follows:

$$\begin{aligned} &I_0^t(Y_1) = I_0^t(Z_1),\ I_n^t(Y_1) = I_n^t(Z_1), \quad t \in I, \\ &(K_t^i(Y_1), C_t^i(Y_1), L_t^i(Y_1)) = \lim_{m\to\infty}(K_t^i(Z_m), C_t^i(Z_m), L_t^i(Z_m)), \\ &\quad i \in I_0^t(Y_1),\ t \in I. \end{aligned} \tag{3.3}$$

It is not difficult to see that the trajectory Y_1 is well defined,

$$Y_1 \in \mathfrak{A} \text{ and } Y_1 \geq Y_0 \geq X. \tag{3.4}$$

We show that property (C2) holds. Assume that

$$Z \in \mathfrak{A} \text{ and } Z \geq Y_1. \tag{3.5}$$

Then

$$Z \geq Y_1 \geq Y_0. \tag{3.6}$$

By (3.1), (3.5), (3.6) and the property (C1) for $t = 0, \cdots, q$,

$$C_t(Y_1) \leq C_t(Z) = C_t(Y_0) \leq C_t(Y_1)$$

and

$$C_t(Z) = C_t(Y_1). \tag{3.7}$$

It follows from (3.2), (3.3), (3.5), (3.6) and the choice of $\{Z_m\}_{m=1}^{\infty}$ that

$$\begin{aligned} C_{q+1}(Y_1) = \lim_{m\to\infty} C_{q+1}(Z_m) = M \\ \geq C_{q+1}(Z) \geq C_{q+1}(Y_1). \end{aligned}$$

Thus

$$C_{q+1}(Y_1) = C_{q+1}(Z)$$

and the property (C2) holds. Lemma 3.1 is proved.

Completion of the proof of Theorem 2.1. Set $Y_0 = X$. Let $\{T_q\}_{q=1}^{\infty}$ be a strictly increasing sequence of natural numbers. Applying Lemma 3.1 by induction for each natural number q we construct a model trajectory $Y_q \in \mathfrak{A}$ such that

$$\begin{aligned} Y_1 \geq X = Y_0, \\ Y_q \geq Y_{q-1} \end{aligned} \tag{3.8}$$

and the following property holds:

(C3) If

$$Z \in \mathfrak{A} \text{ and } Z \geq Y_q,$$

then

$$C_t(Y_q) = C_t(Z),\ t = 0, \cdots, T_q.$$

Using a diagonalization process and extracting a subsequence if necessary, we may assume that

$$I_0^t(Y_q) = I_0^t(Y_1),\ I_n^t(Y_q) = I_n^t(Y_1), \quad t \in I,\ q = 1, 2, \cdots,$$

and there exists

$$\lim_{q\to\infty} (K_t^i(Y_q), C_t^i(Y_q), L_t^i(Y_q))$$

for each $i \in I_0^t(Y_1)$ and all $t \in I$. Define a model trajectory $\tilde{X}$ as follows:

$$\begin{aligned} & I_0^t(\tilde{X}) = I_0^t(Y_1),\ I_n^t(\tilde{X}) = I_n^t(Y_1), \quad t \in I, \\ & (K_t^i(\tilde{X}), C_t^i(\tilde{X}), L_t^i(\tilde{X})) = \lim_{q\to\infty} (K_t^i(Y_q), C_t^i(Y_q), L_t^i(Y_q)), \\ & \quad i \in I_0^t(\tilde{X}),\ t = 0, 1, \cdots. \end{aligned} \tag{3.9}$$

It is not difficult to see that the model trajectory $\tilde{X}$ is well defined and $\tilde{X} \in \mathfrak{A}$. By (3.8) and (3.9),

$$\tilde{X} \geq Y_q,\ q = 0, 1, 2, \cdots. \tag{3.10}$$

In order to complete the proof, it is sufficient to show that $\tilde{X}$ is Pareto optimal.
Assume that

$$Z \in \mathfrak{A} \text{ and } Z \geq \tilde{X}. \tag{3.11}$$

Let q be a natural number. Then it follows from (3.10), (3.11) and the property (C3) that

$$Z \geq Y_q,$$

$$C_q(Z) = C_q(Y_q) \leq C_q(\tilde{X}) \leq C_q(Z).$$

Thus

$$C_q(Z) = C_q(\tilde{X}).$$

Theorem 2.1 is proved.

References

1. V. L. Makarov, *On dynamic models of economy and development of ideas of L.V. Kantorovich*, Economics and Mathematical Methods **23** (1987), 10-24.
2. A. J. Zaslavski, *On a model of economic dynamics with discrete innovations*, Economics and Mathematical Methods **25** (1989), 679-684.
3. ———, *Discrete innovations in dynamic models of economy*, Siberian Math. J. **31** (1990), 47-59.
4. ———, *Optimization of the growth rate in models with discrete innovations*, Siberian Math. J. **32** (1991), 43-50.
5. ———, *Asymptotic behavior of path of a model with discrete innovations*, Siberian Math. J. **33** (1992), 608-617.
6. ———, *On a class of dynamical systems arising in mathematical economics, Part 1*, Dynamic Systems and Applications **7** (1998), 215-236.
7. ———, *On a class of dynamical systems arising in mathematical economics, Part 2*, Dynamic Systems and Applications **9** (2000), 15-36.
8. ———, *Allocations of labour resources on trajectories for the model with discrete innovations*, Advances in Mathematical Economics **2** (2000), 137-160.
9. ———, *Existence of trajectories with unbounded consumption for the Makarov model with discrete innovations*, Nonlinear Analysis: Real World Applications **2** (2001), 195-220.

INDEX

A

B

C

D

E

P

R

S

T

U

V

W